LOCAL

A SEARCH FOR NEARBY NATURE AND WILDNESS

ALASTAIR HUMPHREYS

EYE BOOKS

Published by Eye Books Ltd
29A Barrow Street
Much Wenlock
Shropshire
TF13 6EN

www.eye-books.com

Cover and internal design by Nell Wood
Photographs by Alastair Humphreys
Typeset in 10 / 13pt Adobe Caslon Pro, Shenzhen Industrial and Veneer

British Library Cataloguing in Publication Data
A catalogue record for this book is available from the British Library

ISBN 9781785633676

'And the end of all our exploring
Will be to arrive where we started
And know the place for the first time.'

T. S. Eliot

AUTHOR'S NOTE

'This book will have no scientific value. Those who have studied birds will not find in it anything that they do not already know; those who do not care for birds will not be interested in the subject. The writing of the book, and still more the publishing of it, require some explanation.'

Sir Edward Grey,
birdwatcher and British Foreign Secretary

In this book, I attempt to share what I have learnt about some big issues from a year exploring a small map. Nature loss, pollution, land use and access, agriculture, the food system, rewilding... these are vast, nuanced topics to which people far smarter than me have dedicated entire careers. I've taken an enthusiastic layman's run up at each of them and tried to tick them off in a couple of pages.

Many of these subjects are emotive and some readers will disagree with some of my opinions. I thought it might help if I were to point out that I considered the entire book through a single lens, which I believe unifies all the issues:

We have just this one tiny planet to live on, now and for the foreseeable future. We must care for it, and use its resources wisely, sustainably, and fairly. If we do, then the future is brighter for both humans and nature.

That is my sole agenda and I'm broadly in favour of anything that supports it. It would surprise me to meet anyone who didn't agree with it in principle (if not always in action). So if we differ on specific points, it is better to disagree agreeably and search for common ground where we can make positive changes together, rather than just shouting at each other on social media. I have attempted to acknowledge all sources of information and data at www.alastairhumphreys.com/local. All mistakes are of course my own.

This is an exciting time to be alive. We can become the first generation in history to leave our neighbourhood and the planet in a better condition than we found it, if only we choose to do so. That certainly seems a better legacy than doing nothing, and then earning the contempt of our children and grandchildren.

Alastair Humphreys
Born when the atmosphere was 350 parts per million (ppm) carbon dioxide
May 2023, 422 ppm

A SINGLE MAP

'Instructions for living a life.
Pay attention.
Be astonished.
Tell about it.'

Mary Oliver

For more than twenty years, my favourite thing has been to leave *here* behind, with all its ties and routines. To hit the road and make my way to *there*. I get twitchy being in one place for too long. I have been lucky enough to cycle a lap of the planet, to row and sail the Atlantic, hike across southern India, and trek over Arctic ice and Arabian sands. The open road, spin the globe, and off I go. Home was for family, friends and real life, but not for exploration and adventure.

However, my mood has shifted, like many people's. With the climate in chaos, I can't justify flying all over the globe for fun anymore, burning jet fuel and spewing carbon for selfies. It feels particularly inappropriate as I write books that encourage everyone to get out and explore. If I love wild places so much, was I willing to *not* visit them in order to help protect them?

Flying to distant lands is still a rare luxury on a global scale. Each year, the vast majority of the world's population don't step onto a plane, and just 1 percent of us take more than half of all flights. How can more of us enjoy wild landscapes and the mental and physical benefits of getting out into nature without it costing the earth?

I have been interested in achievable, inclusive adventuring since I began writing about microadventures more than a decade ago, coining that phrase as I encouraged people to undertake weekend bike rides, overnight camps and wild swims. Grand adventures shouldn't just be for people with the time and money to cross continents. Neither should wild places only be for the lucky few with national parks on their doorstep and the freedom to explore that is often affected by gender, race and other factors. Could there be a way to put nearby nature into everyday lives?*

Family life has curtailed my own expeditions in recent years, while of course adding many delights of its own. The merry-go-round of childcare and never-ending chores saw us settled in a less adventurous neck of the woods than I'd ever imagined for myself, on the fringes of a city in an unassuming landscape, pocked by the glow of sodium lights and the rush of busy roads. It is a strange, in-between edgeland: there are fields, but there are factories too. There are villages and farms, train tracks and tower blocks. I don't like where I live. I am here for my family, because they like it and I like them. And that's reason enough. I'd much rather live in their world than live without them in mine. But I had developed a strong tendency to blame the area for most of my frustrations in life, despite being aware of the paradise paradox, which is the belief that moving to a picture-perfect destination will solve all your problems.

It was time for me to accept that we weren't going to move to a croft in the Cairngorms, a cabin in Quebec or a condo in California. But I didn't want that to dampen my enthusiasm for exploration. Could I make exploring my backyard as fulfilling as travelling the world?

One morning I set down the heavy laundry basket on top of the piles of homework scattered over the kitchen table, carried a pair of abandoned cereal bowls to the dishwasher, and looked out of the window.

**I was very proud of coming up with this pithy phrase until my friend Rob Bushby pointed out that I had stolen it from him.*

What if where I live, this bog-standard corner of England, which had held no surprises for me, was actually full of them, if only I bothered to go out and find them? Not known, because not looked for. This was an opportunity to get to know my place for the first time and to search closer to home than ever before for things I've chased around the globe: adventure, nature, wildness, surprises, silence and perspective.

The first step was to get a map. Ordnance Survey, Britain's national mapping agency, divides the whole country into 403 'Explorer' maps on a 1:25,000 scale, meaning that one 1km of land is represented by 4cm of map. You can also order a customised map with your own home right in the middle. I visited the OS website, zoomed in on where I lived, and clicked 'Buy Now'.

I decided to swap dreaming of large adventures for spending an entire year roaming the local map I lived on, an area measuring just 20km across. If you ran across it, it would be shorter than a half marathon. It felt tiny.

A couple of days later, I met the postman at the door and eagerly carried the envelope across the garden to my shed where I write books and plan adventures. There's an old log outside where I could spread out the map. Unfolding a map is the ritual that launches all good journeys.

When the explorers Lewis and Clark set out in 1804, their aim was to survey the 828,000 square miles of the 'Louisiana Territory', stretching from the Mississippi River to the Rocky Mountains, that America had just purchased from France. What was out there? What did their country actually look like, and what opportunities did it offer? I felt a similar call to investigate my own map's more modest span. I wondered what was hiding in plain sight, right under my nose.

I ran my hands over the map to flatten its creases. It was divided into 400 individual grid squares, outlined in light blue, each covering a square kilometre. That's a decent size, about 140 football pitches, but you could comfortably walk the perimeter in an hour. Each week* I would explore one of those squares in depth, doing my best to see

**Or as close to weekly as commitments allowed. I didn't stress too much about the 'rules', as demonstrated by my accidentally visiting fifty-three squares over the year, which I didn't notice until the day this book went to print.*

everything there, to walk or cycle every footpath and street, and to learn as much as I could along the way.

Hand me a map and I'll give you any number of ideas for places to camp and to watch the sunrise, routes to ride, and efficient ways to move from A to B. But I didn't want my habits and confirmation biases to determine where I went this year. I wanted my discoveries to be serendipitous, not governed by my preferences. I was hoping to see things I would not ordinarily see. So, after the first week, I started using an online random-number generator to choose the next square to visit, permitting myself only to veto any squares adjacent to one I had already explored.

From the very first square, it became clear how much of interest there was on my map, so long as I acted on the assumption that everything was interesting. And with that mindset, everything *did* become interesting. The late Sir Terry Pratchett once gave a lecture on 'The Importance of Being Amazed about Absolutely Everything', which felt like a fitting mission statement.

I found myself investigating things I would not ordinarily have noticed: nature in more detail than I'd ever seen it before, and all the history and ephemera I encountered along the way. The amount I had to learn was astonishing. I could tell a daffodil from a daisy, but not a dunnock from a denehole. I'd imagined this would be a year of poking around rabbit holes in the countryside, but it became a year of falling down internet rabbit holes about hundreds of obscure topics, as well as reading dozens of books about history, nature, farming, and the climate emergency.

Anything clever that you read in the following pages, and almost every fact and figure, was new to me when I began this book. Do not make the mistake of thinking I'm a clever person who can stand in an empty field and see biology, geology and every other 'ology, while you merely see a field. I, too, saw only the fields before I started, but paying close attention unveiled so much.

I hope this book encourages you to explore your own neighbourhood, to buy a map of where you live (or borrow one from the library) and use it as a prompt to get active and spend more time outdoors. Share your discoveries with friends and family, observe what you're motivated to learn more about, and then do what you can to get others to care about those things too.*

I don't mention any place names in this book because I want my narrative to be a spark for your own ideas, not a recipe to follow. Discover what surprises are waiting at the end of your own street. Richard Jefferies explored similar liminal landscapes to mine in the 19th century. His book *Nature Near London* also did not specify sites 'because no two persons look at the same thing with the same eyes. To me this spot may be attractive, to you another; a third thinks yonder gnarled oak the most artistic... Everyone must find their own locality.'

Many times during this year of pottering around my local neighbourhood, I thought about Henry David Thoreau, whose book *Walden* is a classic reflection upon simple living in natural surroundings. Appropriately for my project, his cabin was not in the heart of the wild either, but on the edge of a town. He went to live in a cabin in the woods, but could still entertain visitors, go to the shops, and eat pies baked by his mum. Nonetheless, he was very clear about his intentions and they in turn helped to guide my own.

'I went to the woods,' Thoreau explained, 'because I wished to live deliberately, to front only the essential facts of life, and see if I could not learn what it had to teach, and not, when I came to die, discover that I had not lived.'

Finally, in one of my favourite short films on YouTube, *Of Fells and Hills*, runner Rickey Gates pondered something that has stuck with me for years.

'In the end, I think that a single mountain range is enough exploration for an entire lifetime.'

I like that concept. It became the foundational question during this year of trying to live deliberately on my map and to learn what it had to teach. Is a single map enough exploration for an entire lifetime?

**Please share your discoveries on social media using the hashtag #ASingleMap.*

THE DARK

HALF

SAMHAIN

NOVEMBER

BEGINNINGS

'There is in fact a sort of harmony discoverable between the capabilities of the landscape within a circle of ten miles' radius, or the limits of an afternoon walk, and the threescore years and ten of human life. It will never become quite familiar to you.'

Henry David Thoreau

I studied my map for a while and found what appeared to be its most boring grid square. A square without road, house or river, just a single footpath, one pond, and the merest flutter of a lonely contour line. Here was nothing at all, neatly outlined within its crisp blue lines.

It was unremarkable: there was nothing to remark on. It was the ideal place to begin. If this first outing was too boring for me on my boring map, then it would certainly be too boring for you, and that would be the end of the book straightaway.

I folded up the map and headed out to have a look at nothing.

Sometime later, my car was being hauled out of a ditch by two construction workers who were too polite to tell me what a moron I was. I'd flagged down their pick-up to ask for help after my front wheel dropped off the edge of the tarmac into a void hidden by the hedge in the lay-by in which I tried to park. As their engine revved and bits of my car crunched and cracked and fell off, I reasoned I was here to look for new experiences, so perhaps this was a good start?

I thanked the men, pocketed the car keys, squeezed through a barrier designed to keep out dirt bikers, and climbed over a block of graffitied concrete hampering incursions by vehicles or caravans. I'd never been down this way before.

Somebody had planted a row of spindly saplings along the metal fence, tied up by scruffy blue string. Who was it? There were no houses here, and it would be at least a decade before the trees amounted to much. Why had they bothered? I snapped a quick photo.

I paused again just beyond a discarded burger wrapper, this time to admire the colourful, leaf-strewn path. I took a picture of my feet among the golden leaves. The footpath stretched away into the mist beneath a gloomy archway of damp trees. I took a photo of that too. Autumn was making way for winter. It was early November, meaning that the year's cycle had just entered the period known in the ancient Celtic calendar as the dark half of the year. My breath ballooned in the air, and my fingers felt cold even in my gloves.

Celts used to mark the turning of the year with four fire festivals, midway between each equinox and solstice. Samhain was the most important of these, welcoming in the winter and the darker months. People felt anxious at the weakening sun and lit fires to help the sun on its journey across the heavens. These celebrations at the end of harvest were rowdy affairs, filled with gorging, boozing, and sacrificed cattle. Jack-o'-lanterns were carved from turnips and lit from within by glowing coal embers.

Some people took a flame from the community bonfire and carried it home to relight the fire in their own hearth. At this time of year, they believed, the separation between our world and the spirit world dissolved, allowing more interactions between the two. You can imagine their concern at the darkening days, their foreboding about the hungry months to come and the proximity of a closer supernatural realm. The fires and revelry must have been an intoxicating respite.

The customs I was looking forward to on Bonfire Night are not far removed from those pagan rituals whose bonfires and fireworks we borrowed to help us to remember, remember Guy Fawkes and the gunpowder plot. Time stretches back a long way on this small map.

Fireworks were for later, however, and this empty square still seemed unlikely to provide any. I set about trying to find something interesting to look at.

Once I can put a name to something – a bird, a tree – I seem to come across it more often, and I also appreciate it more for knowing the word.* As Robert Macfarlane wrote in *Landmarks*, 'language deficit leads to attention deficit. As we further deplete our ability to name, describe and figure particular aspects of our places, our competence for understanding and imagining possible relationships with non-human nature is correspondingly depleted.'

I intended to make a conscious effort this year to learn more about the nature around me. Addressing my language deficit and general ignorance might help my attention deficit as well, as I seemed to fritter too much of my time on being productive rather than on just being. It was my attempt to get a grip on my nature deficit disorder.

Paying attention is what teachers nagged me to do in boring lessons at school. It was time for me to walk around my map and belatedly learn how to do it. Even though I studied science at A-level and university, I learned little from formal education, and my lack of knowledge was shocking. With these weekly outings came a determination to be more observant, and to fill in some of my knowledge gaps with the help of apps and online research.

My professor today was the Seek app on my phone. It identifies plants and creatures through some unfathomable voodoo magic. I pointed my phone at a common reed, wondering what wisdom the mighty AI gods would bestow upon me. I had seen reeds countless times before but didn't know their official name. Drumroll and revelation, the technical name for the plant was… 'common reed'!

But the breakdown of its taxonomy caught my eye. For this humble reed nestled in the Family of 'Grasses', the Class of 'Monocots', the Kingdom of 'Plants' and the Domain of 'Eukarya'. The sprawling immensity of life, too complex for me ever to grasp, had been ordered and tidied and simplified for this single plant in front of me: *Phragmites australis*. This much I *could* comprehend, and from here I

**This is the Baader-Meinhof phenomenon, a frequency illusion named after a 1970s terrorist group. It is attributed to your brain looking out for the new word and then confirmation bias backing up its apparent prevalence.*

had a foundation upon which I could layer more curiosities.

The open fields of this first grid square were a similar starting point of reference on my little map, in a corner of a country that is forever England, that is in the UK (for now), Europe (ish), on planet Earth in a solar system tucked away in an outer spiral arm of the Milky Way, and on and on until my head explodes into that memorable picture from the Hubble telescope showing 265,000 galaxies (out of billions), many of which are so distant that their light has taken billions of years to reach us. It is a picture sprinkled with trillions of stars and planets and so much untold mystery.

Yet even that photograph has been obliterated by those from the new James Webb telescope, a hundred times more powerful than the Hubble, which is now peering all the way back to GLASS-z13, the oldest galaxy we have ever seen.

So while I stood on a stony path atop deep layers of late Mesozoic and Cenozoic sedimentary rocks, and gawped dumbly at a common reed, I could also open my phone to @nasawebb on Instagram and gawp humbly all the way back to light from GLASS-z13 heading our way a mere 235 million years after the Big Bang when our entire universe (this reed, this planet, this galaxy, *everything*) was as small as an apple and a temperature of a quadrillion – 1,000,000,000,000,000 – degrees!

In other words, faced with the infinite options for exploration out there, I might as well begin right here, right now, on this damp footpath.

This single map contains multitudes. If nothing else felt connected across the individual squares I visited this year, I hoped to hold onto this sudden yawning glimpse of wonder and the connection between an everyday observation and the curiosity that spins off from there if you look at it from a fresh angle or listen to experts.

So now the questions started to come. Who came to this isolated spot to graffiti badly, and why? Who built this bench here from two stumps of birch and a hefty plank? Who laid down a bed of wood chippings around the bench, which was now dotted with crooked brown mushrooms? Who made the effort to gravel a small path over to this bench and cut branches to line the path and peg them down?

I wiped the bench dry with my sleeve, sat down and rummaged for the flask of coffee in my rucksack. I pulled my hood up over my woolly hat and sipped the drink for its warmth rather than the caffeine. My brain was buzzing quite enough as it was. I stared into the damp fog and was struck by another question, more pressing than the others: why had someone put a bench right in front of a tangle of brambles and a massive pylon?

But then I spotted a small plaque, inscribed to Brian, 'a tireless campaigner for the canal', and I realised I was looking at this all wrong. What I should have been looking at from the bench was behind me. I swivelled round to face the other way and then saw things differently. What I'd dismissed as a stagnant ditch was, in fact, an overgrown canal in the early stages of restoration. There were tall bullrushes like hot-dogs on sticks, feathery reeds, blood-red hawthorn berries and a spiked metal security fence.

It was apparent now what I had missed before. The scrub had been cleared to allow a view of the canal from the bench. It wasn't exactly paradise, but this was a framed view of nature, history, conservation, and community all rolled in together.

Brian's bench, and the evident fondness and appreciation for this place that had inspired it, gave me permission to cherish this view too. And as I sipped my coffee, I felt weirdly thrilled to be here on this murky, chilly November morning.

While I was taking all this in, a plump man in his fifties cycled

down the canal path in hi-vis, the tyres on his bike squashed rather flat. Tom Waits' unmistakable gravel voice played from a speaker on the handlebars, and the cyclist sang along as he rode past without spotting me.

'I never saw the morning 'til I stayed up all night.'

I smiled at the break in the silence.

'I never saw my hometown until I stayed away too long.'

There was little movement or birdsong in the air as I finished my coffee, just the sounds of a forklift truck in the industrial yard beyond the canal, reversing beeps, and a rattling train somewhere in the distance. Nature seemed subdued in the morning mist.

It feels ridiculous to admit this, but I still had not actually entered my first grid square. I was foiled from getting into it by a drainage ditch that was too wide for me to risk leaping over in my wellies. I was forced to retrace my steps and to try another way. While walking back down the lane, I picked up the burger wrapper I had ignored earlier. I already cared more for this place.

I pushed through a narrow gap between some bushes and a chain-link fence onto a narrow footpath, taking care not to snag my coat. At last, I was finally into my grid square and ready to begin this challenge! I followed the hemmed-in path until I reached a gap by a fallen fencepost. I stepped over the tangled loops of wire and dropped down a wooded slope to a stagnant green pool. The surface was covered in duckweed, a tiny, quick-spreading plant that I later learnt was being tested in the US as a stage of treatment for human sewage.

It was peaceful down there among the hawthorn bushes; a no-man's-land of Carlsberg Export cans, wedged between a railway line, industrial units, marshland and a Ministry of Defence firing range. Nobody knew I was here. Nobody I knew had ever been here. I didn't know why anyone else would ever have come here, though the crushed beer cans, like proprietorial flags claiming mountain summits, showed I was not stepping into uncharted territory. I stirred the pond with a stick and its mysteries bubbled up from the black depths and stank like a cauldron.

I checked my location using the map app on my phone, picked an apple from a tree by the railway line and popped it into my rucksack for later. Then I climbed into a field and waded through wet, knee-deep grass, among what Seek taught me were yellow common toadflax flowers and purple thistles bejewelled with droplets of dew and strands of silk webs.

I headed towards the grazing meadows of the drained marshland that made up most of today's 'empty' grid square. Gigantic electricity pylons marched across the land and the grey sky was striped with lines of cables running from the old coal-fired power station down on the coast. The building had fallen foul of new environmental laws and the plant was decommissioned, including blowing up the enormous chimney, which must have been a very satisfying morning's work.

The 200-metre-tall chimney gained some notoriety back in the Noughties when protesters climbed it to paint 'GORDON, BIN IT', a message to Prime Minister Brown. But they were served a High Court injunction before they got any further than a colossal 'GORDON'. The activists admitted trying to shut down the station, but argued that their damaging actions were to prevent climate change causing greater damage to the world.

Nobody had used this claim before in a 'lawful excuse' defence. An Inuit leader supported them with evidence that climate change was affecting his way of life. And *The New York Times* featured the eventual lawful-excuse acquittal in its annual list of life-changing influential ideas.

Beneath the pylons, the land was strikingly flat. I could see an elevated rampart far in the distance, beyond this square, protecting the marsh from the wide river that fanned out towards the estuary. A silhouetted pony grazed on the grass bank, and a dirt bike revved up and down it, having somehow found a way around the preventive barriers. The noise, the movement and the human all seemed incongruous on this empty morning as I watched a ship slide down the river behind the rampart, heading out to sea with smoke streaming from its funnel.

'Where are you going?' I called out to the vessel. Which foreign port will you make landfall in? What will you see there? How will it smell? What café will the crew go to for a beer and a smoke to stretch their legs upon arrival? I used to travel to those far-off ports, and contemplating the prospect of being stuck here for a year made me wonder if I was missing out. I continued walking my rough lap of the grid square, heading down towards its third corner.

One feature marked on the map was a tiny mound that had been awarded its own diddy contour ring, a whopping five metres above sea level. It looked nothing more than a grassy wrinkle on the flat counterpane of the marsh. But the internet told me it was an ancient barrow, a Bronze Age burial site constructed over a stone coffin that once contained a crouched skeleton and a necklace of beads made from fossilised sponges. Those stories from thousands of years ago added a sense of awe to the innocuous mound and its empty metal cattle trough.

From my vantage point on the 'hill' I looked over a field of black cows and another of white sheep. The fields were separated by drainage dykes rather than the fences, hedges or walls I was more accustomed to. The livestock were the only clue that this drained marsh was anything other than forgotten ground, in-between ground, left-behind ground. The animals play an important role in maintaining it and preventing it from being engulfed in scrub.

Two crows swooped overhead, calling out as they tumbled and rolled through the cold grey sky. Were they courting? Fighting? Playing? I could hear them swoosh as they dropped. I heard, also, the creaky wings of a pair of lumbering white swans flying by, then the begrudging, cranky take-off of a heron whose frog-hunting I disturbed in a ditch covered in neon-green algae.

I was hungry now, appreciative of the apple I'd pocketed earlier. It was huge, red and flecked with yellow. Lemony sunlight was burning off the morning's mist. The apple's crunch sounded loud in the quietness. I saw the distinctive swoop of a woodpecker and listened for its laughing call. Less easy to identify was either a weasel or a stoat that scurried across my path and disappeared like magic into the long grass. They do say that telling the two apart is simple: a weasel is weasily identifiable, whereas a stoat is stoatally different...

And that brought me back to where I began, the lap of my square complete, but now with a dinged car, a bunch of photographs, and a whirlwind of first impressions. I finished the sweet apple and tossed the core into the hedge. I had pages of notes to begin exploring on the internet when I got back home.

I had selected the most empty-looking location to begin my journey, on a map of an area I'd often dismissed as boring. But what I came away with, knowing that 399 grid squares still potentially awaited, was a sense of abundance and possibility.

There was a clear correlation between how much I had observed today and how much I enjoyed the outing. The government's Foresight programme, which looks at how to improve mental capital and mental well-being, set out five actions to improve personal well-being. These were:

Connect
Be active
Take notice
Keep learning
Give

That is, if you'll allow me to label this book as giving back what I discovered, a perfect summary of what I was hoping I might accomplish in this year – an abundance of new places to connect with, of things to learn, and of beauty to observe, if I kept active and paid sufficient attention. It was a fine beginning.

FLY-TIPPING

'My vicinity affords many good walks; and though for so many years I have walked almost every day, and sometimes for several days together, I have not yet exhausted them. An absolutely new prospect is a great happiness, and I can still get this any afternoon.'

Henry David Thoreau

After the uplifting experience of my first grid square's open spaces, my heart sank a little upon arriving in my second square a week later. I found a boarded-up, closed-down pub with weeds pushing through the cracks in the puddled car park. A soggy planning-application notice on the security fence outlined plans to demolish the pub and build six houses.

Pubs improve community engagement and rank as Britain's third-most-popular tourist activity, yet they are in seemingly terminal decline. This leads to a loss in community fellowship, a sense of belonging that I had not yet found while living in this area, whether down the pub or up the hills. I like pubs in the way that I appreciate churches: not bound to the religion espoused, but enjoying the link to the past, the reassurance of the familiar, and the welcoming open doors.

I thought of the generations of memories from evenings spent in the Anchor and Hope that now awaited demolition along with the building. Its name derived from the Letter to the Hebrews in the Bible. 'We have this as a sure and steadfast anchor of the soul, a hope.' Pub names often give clues to their ages. For example, after Henry VIII's Reformation, many pubs hastily switched from being the Pope's Head to the King's Head.

Dozens of pubs are closing down every week. At this rate, the esteemed British hostelry will be extinct by the 2040s, along with the Sumatran orangutan, Amur leopard, black rhino, hawksbill sea turtle, Sunda tiger and Cross River gorilla. The Covid pandemic sped up the pub closures, which is ironic given that the massive number of pubs in Britain may be connected to the 14th-century plague. So many people died that the survivors found themselves in an emptier land, with higher wages and a greater inclination to have fun.

Leaving the abandoned pub behind, I headed off in search of cheerier discoveries. After last week's parking disaster, I had remembered that bikes nearly always trump cars, and so cycled to today's square, as I would do to almost every other square in the year. Sometimes I would then pedal around the squares, other times I locked up my bike and walked.

On this occasion, I rode down a single-track lane, which was shining from overnight rain as it wound between tight hedgerows. I turned onto a muddy track through a birch wood towards an old gamekeeper's cottage. The thinning leaves were yellowing as winter clamped down. A sign cautioned vehicles to proceed slowly. I didn't seem to be struggling to do that this morning, I was pleased to note. I had travelled about 400 metres in an hour, as I kept stopping to look at stuff.

A channel of trampled grass on the verge caught my eye. I followed it up to a badger sett, where five large holes had been dug into the sticky earth. Badgers are big, handsome and abundant creatures, but very elusive. One dark night last summer, I almost trod on a badger while out running in the woods. It scared the life out of both of us. Later that same week I cycled straight over a fox at midnight – thump, thump went the wheels. The fox fled, so hopefully it was OK. And then the very next night a gigantic, scratchy stag beetle landed on my arm while I was out running, and I leapt out of my skin for a third time. Who needs the Serengeti when you've got the rural urban fringe to explore?

The British badger is a different species from the American badger and from Africa's honey badger, which is more weasel than badger. A quarter of Europe's badgers live in Britain, where they exist uneasily alongside farms, owing to the risk of them transmitting tuberculosis to cows.

Badgers live in setts made up of interlinking tunnels, with tonnes of soil excavated to create dozens of entrance holes and long runs as deep as four metres underground. These complicated structures include ventilation shafts, nurseries, latrines and sleeping chambers. Fresh air circulates as the entrances are higher, allowing stale air to pass out as if through a chimney. Setts are often dug and enlarged by successive generations for a century or more. I peered down the hole with fascination.

The night before, I'd finished reading *The Wooden Horse*, an escape story from the Second World War. Two prisoners of war were contemplating the difficulties of conventional escape tunnels that began inside a hut – beneath a desk, in the showers, under a cooking stove – and which then ran all the way underground beyond the camp's perimeter. The sheer length of the project, in both distance and time, usually gave the German guards too many opportunities to discover the mischief.

What if they'd started their tunnel near the perimeter fence instead? It was a ludicrous idea, but many of life's wonderful projects arise from asking 'what if?' and mulling over madcap possibilities.

The concept that came to Eric Williams and Michael Codner was audacious, with pleasing echoes of Odysseus and his Trojan Horse stunt. They placed a gymnastics horse in the exercise yard, out by the camp fence. While volunteers practised vaulting for hour after hour, a tunneller concealed inside the horse began digging an escape tunnel, concealing it at the end of each day with a sand-covered trapdoor. The conscientious gymnasts tidied away the horse each evening, carrying it back to their hut with the digger and bags of excavated soil hidden inside.

The gymnasts returned to the same spot day after day, showing admirable dedication to their new hobby, while their friend laboured and sweated underground. A fellow prisoner was sure the plan was 'crackers', declaring, 'I give it a couple of days.' But the men persevered, and a small, repeated activity slowly bloomed into something significant.

Looking at the entrance to the badger sett, I tried to imagine myself in an underground space like that every day for three months, using bowls for shovels, digging by candlelight, struggling for breath in a dark 75cm-diameter shaft. The prospect was terrifying, claustrophobia-inducing, and quite mad. Not only that, but once the escapees completed the 30-metre tunnel and broke out, they still had to cross 150 miles of enemy territory in homemade disguises, without being able to speak German, and then sneak onto a ship to Sweden. The convoluted artifice of modern adventures would seem risible to those bold, courageous men.

As I mulled over the change in direction of my adventures, I was fascinated by the adventurous spirit and focused purpose of those prisoners of war set on escape and freedom. I was intrigued as to how that generation managed to return to normal life after their heightened experiences of war. Perhaps because the 'excitement' had been so extreme, greater than any mere expedition, they felt satiated and eager to settle down, rather than left with unfinished business and an uneasy sense that a single map may never be enough?

I was maybe taking the 'Slow' sign too literally today, for I had still barely made it down one flank of the grid square and it was almost lunchtime. I rode through a coppice glowing with golden sunshine, back to the road. That the trees had been coppiced told me they were hornbeam not beech, for beech is rarely coppiced because of its dense canopy. I often muddle up the two, though I ought not to because beech has glossy leaves with small 'teeth', while hornbeam's leaves are rougher and their serrations alternate between large and small 'teeth'.

The only house along the lane was built with brick interspersed with decorative flint motifs. Until bricks became cheap, a century or two ago, most homes and churches in this area were built with flint. It is found as rounded nodules across chalk landscapes, for it originated from sponges in Cretaceous seas, millions of years ago. The sponges extracted silica from the seawater for their skeletons, and when they decomposed the silica accumulated on the seabed and in the burrows of urchins and worms. Flint's nodular shapes often reflect the shape of the burrows it formed in so long ago.

I took a footpath behind the house and pushed my bike across muddy, undulating fields, past cross-country horse jumps, past a pair of dog-walking mums in wellies and sunglasses, sipping from Thermos mugs and discussing their children's teacher. An unseen man beyond a hedge sneezed and I called out, 'Bless you!'

This was countryside put to use: for work, for living, for strolling. It was different from last week's deserted marsh, which I preferred even though this landscape was more traditionally 'beautiful'. This project was already beginning to challenge my assumptions of what was beautiful or natural in the landscape.

On I went, past a plastic lid in the mud with 'Strawberries and Cream (milk)' written on it in marker pen; past blooms of yellow hawkweed, which Pliny, the Roman author, naturalist and philosopher, believed hawks ate to strengthen their eyesight; past holly saplings sheathed in plastic tubes; past wild bees buzzing from their nest in a hole in an ash tree…

Past a plump little nuthatch pecking a branch as autumn leaves fell from the tree, slower than rain.

Past a snail that I rescued from its perilous road-crossing mission.

Past all these things that distracted me, drew me in, and would send me investigating hither and thither across the internet when I got back home.

My head was saying 'Hurry'. My heart was asking, 'Why should I?'

All these diversions guided me slowly towards the end of the grid square, and a failed attempt to investigate a small pond marked on the

map but sealed off by a high fence. I was about to give up and go home when piles of roadside rubbish caught my eye, dumped along the access road to a Gypsy and Traveller site. This was fly-tipping on an epic scale.

Curiosity drew me along the lane lined with heaps of litter and culminating in a smouldering mountain of soggy trash, maybe 20 metres long and a few metres high, including upturned sofas and builders' rubble sacks. Some of it had been partially burnt. Two workmen were surveying the mess.

'The fire brigade has just been to put it out,' one told me. 'Now we've got to get rid of it all. Again.'

The council hired them, I learnt, to manage various notorious fly-tipping sites in the area. They had driven for an hour to come and sort out the eyesore. Two men with two flatbed trucks, waiting for a third colleague to arrive with a digger to scoop up someone else's rubbish: it was an expensive morning for the local council. Britain spends more than £67 million a year cleaning up public land and prosecuting the few people caught fly-tipping.

'Have you been here before?' I asked.

'This is the third time we've cleared this site in the past two weeks. It's getting silly. Sometimes they even dump a new load while we're still here clearing up the last lot. They honestly don't care. There's nothing I can do about it.'

'Who does it?' I asked.

'Who do you think?' he replied, nodding his head up the road.

'I've no idea,' I said. 'At least it keeps you in a busy job?'

'Nah, but it's not right, is it?'

A car pulled up alongside us and two young women from the council got out. They had a weary, not-this-again look on their faces. They began to confer with the workmen about how to deal with this latest episode. I wished them luck and carried on.

Two Traveller children from the community at the end of the lane approached me. The girl was thumbing through her phone and the boy twirled a catapult round his finger. They looked to be about twelve.

'Did you make that?' I asked, nodding towards the boy's catapult. It had been carved from a forked stick, with the bark peeled off and the wood polished with varnish. Neatly whipped twine held the elastic in place.

'Yeah, I made it,' he said, hesitantly, though with a hint of pride. I

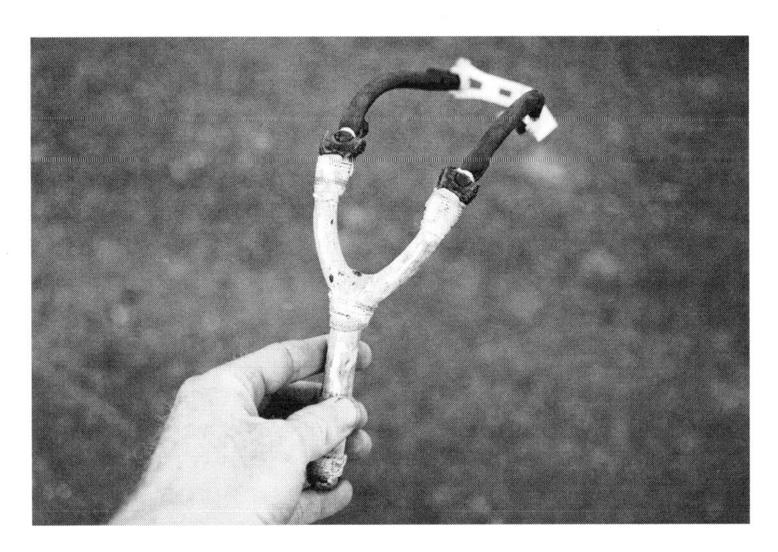

reached out my hand and he let me hold it.

'It's really good,' I said. 'I like it.'

As luck would have it, I had a handful of clay catapult pellets in my pocket (I have a catapult in my shed as a handy procrastination toy) and I offered them to him. He beamed and let me have a shot with his catapult. I took aim at a fly-tipped fridge and missed. Then the boy blasted some dumped flowerpots. All this rubbish had its advantages.

'What you doing with that big camera?' the girl asked, looking up from her phone for the first time.

'I'm trying to see more of the countryside round here, like that pond over there.'

'It's private,' she said. 'You'll get a twenty grand fine.'

'That's a bit expensive for a photo of a pond,' I replied. 'I think I'll just leave it.'

'But that's only for kids, to stop us going in there and messing about,' reassured the boy. 'You'll be all right.'

The pair were warming to me and told me a way to sneak through the fence to the pond. They offered to take me to a nearby graveyard or show me a field of their ponies. They were proud of their home and wanted to show me around, but it didn't feel right to go off exploring with two children, so I demurred.

'Why aren't you in school?' I asked, prompting only a shoulder shrug from the girl and a flurry of exploding flowerpots from the boy.

'Can I have a try with your camera?' he asked.

I showed him how to use the viewfinder and zoom, then posed for him to take a couple of shots of me.

'If they run off with your camera, I'm not chasing them,' shouted one workman, laughing.

'Don't worry. I back myself over fifty metres,' I replied, smiling at the boy. They were nice kids.

But the woman from the council called me over and queried why I was taking photos.

'Be careful,' she said with concern in her voice. 'The people who live here won't like seeing you with a camera. The kids don't ever speak to me; they just tell me to eff off.'

My time was up for the day, anyway. I needed to hurry home to collect my children from school. I shouldered my camera and turned to leave.

'Bye guys,' I said to the kids.

'See you,' they replied with a smile. 'Thanks for the catapult pellets.'

DENEHOLES

'I am often asked what luxury item I carry with me, and the answer is always the same – it is the thing that I left behind.'

Ray Mears

I'd ridden to today's square straight after waking up, so still felt morning fuzzy as I sat on a log to settle into the woodland and sip coffee from my flask. The ground in the hollow was covered with a carpet of leaves, brown beech, tawny oak, orange sweet chestnut, yellow maple, and emerald moss. A brilliant blue sky framed the few leaves still lingering on the branches of silver birch trees. This was likely to be the one of the last days of late autumn colour, making it even more precious. Winter's onslaught was being held back by these occasional autumnal flurries, but there would not be many more golden mornings until next year.

I spent a happy half-hour trying to perfect an artistic photograph of a crooked oak tree in a sunlit patch of bracken, trying to get that sparkling lens flare you can conjure with a wide lens and a small aperture. I hoped the tree might be home to the two species of rare oak-munching beetles whose presence had secured this wood's status as a Site of

Special Scientific Interest (SSSI).*

Most of the photos I took today, however, were not quaint autumnal scenes, but images of burnt-out cars or the detritus of clay pigeon shooting, thousands of smashed orange clays and mounds of plastic shotgun cartridges. There were more wrecked cars than wrens in this wood, rusted and graffitied, the windows smashed, tyres and interiors burnt away, abandoned by joyriders after a night's entertainment ragging around the nearby city. I got my own small thrill by jumping up onto the roof of one to snap a self-timer picture. Then I hid my bike in some trees and stepped away from the paths and the scrunched-up cider cans peppered with bullet holes.

I pushed through the foliage into a thicket sunk in a dip and tangled with fallen, moss-covered trunks. Stringy ash saplings rushed for the sky, charging for the light. It was a literal race of life or death, for only one or two of all those young trees would survive in the long run. As with the stagnant pond from week one, I enjoyed the hidden, unkempt atmosphere of the glade. But it also saddened me that on my map, a mere scrap of undisturbed woodland felt wild and expansive.

Shifting baseline syndrome is the way change happens so slowly that we don't recognise our perception of 'normal' changing. I think I look the same in the mirror each morning until one day I can't button up my jeans. We don't realise a landscape is deteriorating until it's too late. It is one of the most important concepts I learnt all year.

Some scientists believe shifting baseline syndrome is the biggest challenge in conservation today, warning that 'our conservation and policy efforts are becoming less effective with each generation, while we become more satisfied with our diminishing actions because our targets are weaker in terms of biodiversity and habitat variety'.

Three hundred years ago, for example, there were just 6 million people and as many as half a billion birds in England. Today there are nine times more people and a third the number of wild birds (one in six birds has disappeared in my lifetime alone). So much birdsong and space would blow our unaccustomed ears and minds. It is heartbreaking when we stop to consider what a drab state of nature-deprivation we exist in. But we consider it normal and therefore what should be preserved.

**The shimmering two-spotted oak buprestid and the oak pinhole borer, since you ask.*

I would love to have stood in a field in North America and watched the mile-wide flocks of migrating passenger pigeons that roared like thunder and blocked the sun for three straight days as they passed overhead. It was, perhaps, the world's most abundant bird. But the last individual of that species, named Martha, died in a zoo in 1914 with a palsy that made her tremble. And that was it. We had hunted the passenger pigeon to extinction and none of us born since then has even missed those birds, for we live under a new normal.

Near my home there is a wood I often run through. One day, however, I found a tall metal fence blocking the path, shiny new with sharp spikes on top. I had no option but to turn around, furious. How dare anyone block a harmless path through the trees? That patch of woodland had been stolen from me, and diminished my world as a result.

Now imagine a newcomer to the area. They saw me galloping heroically across the landscape and were sufficiently impressed by my unflattering Lycra, sweaty red face and lumbering stride to also go for a jog through the wood. They would reach the fence and just follow it around, accepting it, and still enjoying their run. Their baseline for the natural state of that woodland is lower than mine. Shifting baselines lower standards imperceptibly but relentlessly, so we don't pay attention to the catastrophe of environmental degradation and loss of wild places.

Baselines also complicate our relationship with land use. For example, the purpose of our national parks is to 'conserve and enhance their natural beauty, wildlife and cultural heritage'. This sounds admirable. But their baseline assumptions sit from when they were founded, in 1951. Ought we to conserve those landscapes as they were then? They already looked very different to how they had in 1551, say, or 1951 BC. We assume we have always intensively farmed our countryside as we do today. Maintaining that status quo therefore often falls under the banner of 'conservation', despite its many harmful impacts on nature. Asking what is the 'right' condition for a landscape would become a regular conundrum as I explored my map.

The atmosphere felt timeless in the glade where I stood, but in reality no tree here was more than a couple of hundred years old. It was an evolving and lived-in landscape. Once upon a time, the Romans built a road and settlement just down the hill from here. But by the time they turned up with their pizzas and noisy Vespas, this area had already been managed and shaped for millennia.

The wood was still surrounded by a Medieval embankment, originally constructed to keep out livestock and protect the wood. Over the ages, its timber has been used for construction and tools, in households and for making charcoal. Early industries, such as ironworks, lime kilns, potteries and brickworks were often established in woodland because of the abundance of fuel. Quarries were dug in them, too, rather than losing precious agricultural ground, hence today's overgrown quarry now covered with exploded clay pigeons.

Wherever I walked on my map, I trod on ancient history and cyclical patterns of people using the land, moving on, and then nature reclaiming and rewilding it. This actually made me feel hopeful, for if we don't plunder it, nature's powers of regeneration and renewal are astonishing.

I thought I had done a decent job of exploring today's square until I came home, researched the earthwork boundary, and learnt that the wood contained a 'denehole', something I had never even heard of and must have missed. After every outing, I will spend hours online learning about what I've seen, before turning my thoughts to the next week. But, on this occasion, I realised I needed to return to my square, and this time with a climbing rope and a headtorch…

Deneholes are man-made, underground caves accessed by a vertical shaft. Chalky landscapes were once riddled with them. Theories used to abound about their history – druid temples and elaborate animal traps among them. In reality they were nothing more than tiny mines, sunk to gather chalk to spread on fields as fertiliser. Digging down before excavating the chambers avoided wasting farmland and prevented the pit from filling with leaves.

Deneholes were common in the Middle Ages, though some are so old that they were excavated with bone picks. Pliny described the British digging them as long ago as AD 70, and in 1225 Henry III gave anyone the right to sink a 'marl pit' on their land.

The entrance I found was a small black hole, just a few feet wide, on ground otherwise thick with leaves. I forgave myself for missing it on my first visit. Originally, the entrance for this 800-year-old denehole was a six-metre vertical shaft, but a grate now covered it to prevent accidents. However, at some point, one of the pit's chambers had

collapsed, meaning I could slither down through a slanting hole in its roof instead.

I texted a friend my what3words* location in case I did not reappear and needed rescuing, lashed my rope to a tree, then leant back into learning something new. I switched on my head torch and then lowered myself hand over hand into the hole. This was more excitement than I'd imagined my map would offer within earshot of a busy motorway.

I am claustrophobic, so squeezing through the narrow gap gave me more cause for concern than I'd expect to feel in a small suburban wood. Thankfully, the tunnel opened out after a few metres into a lofty chamber with white chalk walls. It was a humbling feeling to cast my thin beam of torchlight around the pitch darkness and so much history.

I looked up the vertical entrance shaft, capped with its grille. I could faintly hear traffic and the 21st century, but I felt very far from the world. I was glad my friend was awaiting a text message to confirm I got out OK. It would take a long time for anyone to find me if I keeled over in this ancient chamber.

The underground excavations were larger than I'd anticipated. There were six domed chambers, each measuring about three by four metres, with the ceiling arching five metres above me. It had taken a

* *what3words is a location system that is easier to remember than latitude and longitude, simpler than grid references and more accurate than postcodes. It's a very handy app to have on your phone.*

ferocious amount of work to extract all that chalk with primitive tools and lighting. In my torch beam, I saw that visitors across the centuries had etched the walls with graffiti. The plastic bottles lying around testified to more recent explorers.

Suddenly, my heart gave a jolt and I swept the light back to see what had caught my eye. A skull! The darkness, the bars over the entrance… My imagination leapt into overdrive.

I stumbled over the rubble to look. With a mixture of relief and disappointment, I realised it was a cow's skull, rather than anything more historical or horrific. Quite how it ended up down there, I preferred not to know.

I shone my torch around for a while, admiring the endeavours of those long-gone farmers and enjoying the novelty of my morning. Then, huffing and puffing and hauling on the rope, I wriggled back up through the tunnel into the daylight.

I was buzzing to have discovered this denehole just a few miles from my house. It was one of the most interesting things I had seen in Britain, and yet I'd never even heard of these places before. What other surprises might be hiding on my little map?

GROWING

'I felt like lying down by the side of the trail and remembering it all. The woods do that to you.'

Jack Kerouac

A sweaty man ran towards me, suffering the exertions of his morning run. This was the first grid square I'd been to that was predominantly residential. I stepped aside on the church path to let him pass, but he just touched the trunk of the yew tree in front of me and ran back the way he'd come. The tree must have been his regular turnaround spot. I wondered if he gave it any thought, or if he was focusing more on his split times and Strava ranking.

If I was the sort of man who had a Top Ten ranking of trees in my head (and I *am* that sort), then I'd rank the yew far down the list. They look gloomy, suck away sunshine, and are hard to climb.

Perhaps I am being unfair. Yews live for an impressively long time. The Fortingall Yew in Perthshire is 2,000 years old, and the Defynnog Yew in Powys may be 3,000 years older than that. They are among the oldest living things in Europe, but have less legal protection than half a million of our listed structures, which include a few bus stops and skate

parks. Britain has the most yews over 500 years old in Europe. We can't often make that sort of claim. The Ancient Yew Group has identified well over a thousand here. Compare that with France's paltry seventy-seven or Spain and Germany's risible four each.

Long before the development of the Christian tradition of planting yew trees in graveyards, the tree was sacred to druids. They revered its longevity and powers of regeneration, for when its branches droop to the ground they can themselves take root and form new trunks. The yew came to represent both death and resurrection, connected perhaps to the notorious toxicity of the yew's needles. The deadly brew that Macbeth's witches concocted included 'slips of yew, silvered in the moon's eclipse'. Medieval longbows were also made from yew and were used with shivering effectiveness in the Hundred Years War, raining down ten aimed shots per minute upon their foes.

Closing the church gate behind me, I headed to the village green, where I came across the usual war memorial for the usual dozens of young men from this village killed in the First World War. Abbott, Ashdown, Baldwin, Ballard, I read. Beal, Beckett, Blunden… I wondered if there was a single community on my map without such a poignant memorial. It seemed unlikely, sadly. Villages where all residents survived the First World War are known as 'Thankful Villages', and there are only fifty-three in England and Wales, out of tens of thousands. France has just one such Thankful Village, Thierville in

Upper Normandy.

I was feeling a little melancholy after the graveyard and the memorial, but the bus shelter cheered me up. Someone had donated a pile of books to help pass the time waiting for the hopelessly infrequent bus service, with rural public transport being virtually non-existent these days. I like a pop-up library.

Opposite the shelter was a pub and a charming timber-framed home, a yeoman's hall, dating from the 15th century, with high ceilings, and a hearth in the centre of a large dining hall. It would have been built by a successful farmer who owned his own land, known as a 'yeoman'. What would he have made of the food on offer up the street from his house today? The Indian takeaway's 'Tiffin Box Meal Deals' seemed a long way from medieval pottage, a thick stew made by boiling vegetables, grains and any available meat or fish. Tiffin boxes are a stack of three or four metal containers, like a fancy packed-lunch box, popular in India, though I suspect this takeaway would use the usual disposable plastic tubs.

Mumbai has an industry of *dabbawallas*, workers who transport more than 130,000 of these tiffin lunchboxes across the city every working day, because Indians won't put up with a soggy, clingfilmed cheese sandwich in the office.

Every morning a *dabbawalla* collects a tiffin box, known as a *dabba*, filled with fresh food, from the customer's home in the suburbs and takes it to the local railway station for delivery into the city. There the *dabba* is handed to another worker, who delivers it to the correct office at lunchtime. In the afternoon, the process runs in reverse, returning the *dabba* to the customer's house so they don't have to carry their own lunchbox home.

This 130-year-old business has become famous for its astounding accuracy. What makes the system more impressive than simply its scale and efficiency (and fascinating to distribution companies such as FedEx) is that the semiliterate *dabbawallas* accomplish these incredible logistical feats with no phones or computers.

From the village green with its old church, traditional pub and solid homes, I moved outwards through rings of newer housing estates, built one by one over the past decades, each in a distinct style. I passed maisonettes, blocks of flats, cookie-cutter family homes, a house-proud

garden next to a tangle of weeds with an old car up on bricks, and the home of Sheila and Malc, whose names were carved proudly on a wooden arch in the small but flower-filled garden of their terraced bungalow. A happy home, I guessed.

These residential rings around the old village had swollen over time, like galls on an oak tree. I enjoyed ambling through the different areas and wondering at the variety of families and lives and stories there. I felt a sense of 'sonder'. This is a German word meaning 'special', but it is gradually being absorbed into English. The Dictionary of Obscure Sorrows website, which coins neologisms for emotions that do not yet have a specific term, define 'sonder' as a noun meaning 'the realisation that each random passerby is living a life as vivid and complex as your own, populated with their own ambitions, friends, routines, worries and inherited craziness'.

Many streets were named after birds, flowers and trees, a reminder that, not so long ago, all this was countryside. Today I watched a man mowing his lawn five weeks before Christmas, shaving as much nature as possible from his personal square of greenery.

Incongruous among all the new homes was a tiny lodge house, an old single-storey sandstone building with two windows, a door between them, and a triangular roof. Once upon a time it stood at the gates of the long driveway to a Tudor manor house, the grounds of which were now the focus of arguments over 400 proposed new houses. As I prepared to take a photo, a white van braked hard on the road.

'Don't take fucking photos!' shouted the driver.

'Why not?' I asked.

'Because it's my house,' he lied, cackled, and roared off with a wheelspin.

I shrugged and took my photo.

Much of this grid square was covered with orchards belonging to a fruit research station, kept behind fences out of concerns for biosecurity. Growing an apple tree isn't as simple as just spitting out a pip; you often end up with a crabapple tree if you plant an apple seed.

Though that curmudgeonly tough guy Thoreau claimed to prefer wild apples 'of spirited flavour', even he had to admit that sometimes they were 'sour enough to set a squirrel's teeth on edge and make a jay scream'.

Modern apples are hybrids of wild apples, facilitated by trade along the Silk Road over many centuries. Indeed, the etymology of 'Almaty' in Kazakhstan may derive from a phrase meaning 'father of apples'.

Apples evolved to tempt large animals to eat them, so the apple you enjoy for lunch (after your soggy cheese sandwich) came about through the efforts of extinct megafaunal herbivores, Silk Road merchants, plus the boffins hard at work in my grid square. The apples we eat today are cloned by grafting, and this research station produces rootstocks that are used in 90 percent of Europe's orchards.

I sat outside the orchard and ate my Ecuadorean banana, looking up at a silver maple tree draped with large clumps of mistletoe. Cultivated apple is mistletoe's favourite host, and it is spreading eastwards through Britain, perhaps because more blackcaps now spend the winter here rather than migrating to warmer climes. The birds eat the white berry flesh then wipe their beaks on branches to leave behind the seed. This helps the mistletoe to germinate and take hold on new trees.

The Christmas tradition of kissing under sprigs of mistletoe may have originated with Celtic druids almost 2,000 years ago. The plant was a sacred symbol of life, as it blossomed even in the bleak midwinter, and was used on both humans and animals to restore fertility. However, the first record of mistletoe kisses comes only from a song published in 1784. The original custom decreed that you had to pick a berry from the sprig before taking your kiss: once the berries were all plucked, that was the end of the canoodling.

As the research orchards were off limits, there wasn't much else to investigate. I was delighted to see my first fieldfares of the season, and I added 'Large Fries' to my growing menu of McDonald's litter. I spotted the first fish of my map, too, a lively school of dace sheltering under a bridge in a stream.

Small brown trout also nosed into the current, hovering with a gentle flickering of their tails, camouflaged with golden flanks and dark spots. The brown trout is a fierce predator of smaller fish and insects. It cheered me, after the day's convenience stores filled with packaged food, and orderly orchards, to know that, even here, I was in the company of 'fierce predators'.

Nature is all around us. We are in nature. We are nature.

DECEMBER

FOOD FOR THOUGHT

'When you change the way you look at things, the things you look at change.'

Albert Einstein

As I cycled to today's allocated grid square down country roads, over a dual carriageway, around a housing estate, through some villages, past an orchard, a quarry, and lots of farmed fields, I pondered this ordinary mix of features I generally took for granted on my map. How does it compare to yours? How has it changed over time? And what impact does the way we use land in our neighbourhoods have on the planet?

It is hard to grasp the proportions of things that make up our country, and I hoped bumbling around my map would help. The official breakdown is that 28 percent of Britain is pasture, 26 percent arable, 9 percent peat bog, 7 percent moorland, 6 percent grassland, 5 percent coniferous forest, 4.9 percent homes and gardens, and so on.

On a global scale, the numbers are quite different. Seventy-one percent of the planet's surface is ocean for starters, not particularly habitable unless you're in a rowing boat with a supply of dehydrated food, and – having given that a try for 3,000 miles – I can't recommend it as a barrel of laughs. That leaves 29 percent of the planet, of which almost a third is glaciers and barren land such as deserts, salt flats, rocks and beaches. The remaining area is all that we have to live on and share with nature.

Only 1 percent of this area is built-up housing and infrastructure, a statistic that filled me with optimism. I often feel despondent that the world is disappearing beneath a crust of concrete. But on a global scale, 8 billion of us take up a reassuringly small amount of space.* Urban living is also, potentially, the most environmentally friendly way for most of us to live as it requires fewer resources and less power and space per person.

Freshwater takes up another 1 percent of the space, 14 percent is shrubland and 38 percent is forest. But I was stunned to learn that the remaining 46 percent is used for agriculture. We use almost half of the only known habitable land in the universe to make our lunch! My optimism about humans not taking up much space was crushed.

Food production not only takes up a staggeringly vast proportion of the earth, but it also exacts a devastating toll. What we eat and how we produce it is integral to climate change, water and pollution crises, and the runaway destruction of nature. I was astonished to learn all this, and then did my best to ignore it for as long as possible. I had to read a lot of books about farming, food, and climate change before I was reluctantly compelled to address the contents of my supermarket trolley.

To say there are terrible problems with our food and farming systems is not a complaint against farmers, especially not against nature-friendly farms using methods that champion sustainability and biodiversity. But farming has industrialised since the Second World War. The green revolution dramatically increased yields and the world now produces enough calories to feed 10 billion people.

Farms had to become ever more efficient to stay afloat, at the mercy of government policies and competitive supermarket contracts. Supermarkets demand low prices from farmers, and we consumers cheerfully gobble up the cheap food with little concern for, or

**World population is set to peak at around 10.4 billion in 2086, before declining slowly to a new equilibrium. Population size alone is not going to be the long-term catastrophe some fear it to be.*

understanding of, how unsustainable industrial agriculture has become.[*]

To try to keep vaguely within the themes of this book, I'll leave aside the weird ingredients in processed food,[**] the health crisis of excessive cheap calories,[†] and the cruelty involved in industrial animal farming.[‡] I will stick to just three ways in which farming directly affects nature and the planet: emissions, pollution, and land use.[•]

Agriculture and the food system account for more than a fifth of all greenhouse gas emissions responsible for global warming. While there are many benefits to eating local food, what we eat dictates our environmental food footprint far more than where it comes from. Meat and dairy products tend to emit more greenhouse gases than plant-based food, with beef, lamb, cheese and milk being the worst offenders – by far – per unit of protein produced.

British cattle emissions are below the global average per cow, but the methane emissions of the world's fifteen largest meat and dairy companies are higher than those of entire countries such as Canada or Australia, and equivalent to more than 80 percent of the EU's entire methane footprint.[††]

[*] *British families have gone from spending 35 percent of household income on food in 1945 to just 10 percent today (though the lowest-income families pay proportionately more), while the number of calories consumed worldwide has increased. Britain is the most overweight country in Europe, we eat up to 50 percent more calories than we realise, and 80 percent of adults are predicted to be overweight or obese by 2060.*

[**] *The xanthan gum in ice cream comes from a slime that bacteria produce to allow them to cling to surfaces, etc.*

[†] *Type 2 diabetes, colon cancer, cardiac, respiratory and liver diseases etc.*

[‡] *China's twenty-six-storey pig skyscrapers slaughtering a million pigs a year, the British poultry farm with 1.7 million birds crammed inside, the egg industry gassing or crushing unneeded male chicks, etc.*

[•] *I don't live on the coast, so there are no marine explorations on my map, but eating seafood is also often terrible for the environment, with overfishing being a critical problem, fish farms causing pollution, trawling for prawns generating wasteful bycatch, and 92 percent of the oceans being unprotected.*

[††] *Today, 60 percent of the mammals on earth, by weight, are livestock. Humans account for 36 percent and wild mammals make up just 4 percent. The number of chickens has trebled since 1990, to more than 34 billion. Our human population is growing at 1 percent a year, while livestock is increasing at 2.4 percent. Global average meat consumption per person is 43kg a year, and rising towards Britain's 82kg, which is equivalent to eating me once a year!*

Industrial agriculture is also responsible for loss of habitat, soil degradation and the pollution of 60 percent of Britain's failing rivers with animal effluent and chemicals. Take the river Wye as a single example: two-thirds of it faces ecological crisis caused by run-off from 44 million chickens crammed into sheds along the river. Its once pristine clear water, home to fish and wildlife, and held dear by swimmers and kayakers, is now little more than a slimy drain.

Our current global food system takes up an area of land equivalent to all the Americas plus China plus South-East Asia. It is no wonder there is little space for wildlife. We would need to farm double that area if the whole world adopted a British diet, and the planet is literally not big enough to provide everyone with an American diet.

Of all this vast area, 80 percent is farmed for beef and dairy production, much of it via the inefficient process of growing crops to feed to cattle, which then feed us. This is also the biggest global contributor to deforestation and habitat loss.* If everyone became vegan, it could free up an area of land the size of the USA, China, Australia and Europe combined, feeding the world while also reducing farmland by up to 75 percent. That is an astounding amount of space that could be given back to nature and carbon-absorbing wilderness.

The scenario is hypothetical, of course, and there's plenty of small print for people to argue over about farming animals.** But perhaps the sheer scale of land currently devoted to livestock, on top of all their greenhouse gases and pollution, could be food for thought the next time we contemplate a cheeseburger.

To fix the environmental disaster of industrial farming we need to do three things: change what we eat, change how we farm, and change the way we support farmers.

I'd begun today's grid square by nosing around a farm shop selling 'Pasture for Life' meat. The Pasture-Fed Livestock Association promotes a system whereby animals eat only grass and forage crops,

**The expansion of pastureland to raise cattle is responsible for 41 percent of tropical deforestation.*

***Issues of fertiliser use and distribution, local employment, the importance of a single cow for subsistence farmers, the benefits of grazing winter cover crops on arable fields, the biodiversity of having both grasslands and woodlands in rewilding areas and animals' role in that, etc.*

producing meat that is healthier, tastier and generally less harmful to the planet than the more familiar meat labelled 'grass-fed', which means only that the animals eat some grass, but could spend most of their life eating cereals, or soy grown on deforested rainforest land. The grassland on Pasture for Life farms is important for capturing and storing carbon and can be very biodiverse.

If we are serious about sticking to our nation's pledges to get to net zero by 2050 and protect 30 percent of the UK's land by 2030, it is clear we have to consume significantly less meat and dairy produce.* They must become occasional and expensive treats, bought only from regenerative farms and those that improve ecosystems and biodiversity, such as those accredited by Pasture for Life.

This shift in meat's provenance will be harder for the poorest in society, as the farm shop's mouthwatering ribeye steak – red, marbled and delicious – cost an eye-watering £35 per kilogram. Compare this with £14.62 per kilogram at Lidl or £18.73 at Tesco and you see the impossibility of farming more sustainable meat without pricing out most people most of the time.**

Altering our diet will lower emissions, reduce pollution and deforestation, and free up land for wildlife and planting trees, while also

**Indians eat less than 4kg of meat per capita per year, and it would be a brave Brit who claimed our food was more delicious than theirs.*

*** As a loose comparison, you'd get the same amount of protein from £5.20's worth of fava beans, although that might be quite a gassy feast.*

producing more food in our country and protecting our food security. These are all positive things, but appealing to people's goodwill alone won't change enough, quickly enough. As long as industrial farming is allowed to produce cheap food with no punishment for wrecking the environment, there is little hope for shifting consumer behaviour. The hidden costs of all the damage – 'negative externalities', in economics jargon – need to be factored into food prices as a sin tax to repair the mess and speed up change.

The second thing we must do is to change the way we farm, undoing the mistakes of recent decades. James Rebanks writes passionately about both his farming life and his love of nature in *English Pastoral.* His grandfather taught him to work the land the old way, but he has witnessed how much has gone wrong since then. He warns that 'the current economics of farming are such that almost no genuinely sustainable farming is profitable at present. Farming for nature is economic suicide.' Many farmers, like Rebanks, want to be commercially viable while also caring for the land they treasure. We have to help them to move towards food being a byproduct of conservation, rather than an apocalyptic obliterator of nature. Rebanks points the finger at us as being part of the problem, with our demand for cheap food and being 'strangers to the fields that feed us'. My weekly outings are an attempt to change that dislocation in myself. Of course, my observations only scratch the surface and there is no single solution.

Farms need to start producing food in sustainable ways, ranging from intensive greenhouses to nature-led upland farms with low livestock densities. The key is to use each area of land in the best specific way possible. That means apples here and wheat there. It means rewilding here and organic there.

Regenerative agriculture that heals soil, cleans water, and increases biodiversity is vital for our future. It swaps ploughing for direct drilling and uses cover crops, crop rotation and mob grazing (short-duration grazing with extended recovery periods) to care for the soil.* The good news for the vegan-haters** is that livestock plays a small but important

**The Pontbren Project in Wales is a good example of farmer-led regeneration that improves both business and nature. It takes an innovative approach to woodland management and tree planting to improve the efficiency of upland farming.*

***Being vegan is a huge missed steak.*

role in this future of farming. As the regenerative agriculture slogan goes, 'It's not the cow, it's the how.'

Where farmland isn't suitable for growing plants for people to eat, low levels of livestock can help with agroforestry that combines trees with crops and animals, with rotational farming and fertilising, and to help increase biodiversity in habitats. The wildflower meadows we treasure but have almost completely lost, for example, depend upon low-density, occasional grazing from livestock. Imagining all these positive futures for farming makes me very excited.

The third change needed is to support farmers properly in their transition to sustainable agriculture. Subsidies have long been allocated according to land area or number of animals, regardless of their environmental damage. It makes more sense to back farmers who fix nature rather than wreck it, yet the world still subsidises harmful agriculture and fossil-fuel industries to the tune of £1.3 trillion each year.

Farmers are the custodians of our countryside, responsible for far more land than our national parks or nature reserves. But they need meaningful backing if we want them to grow food, reduce emissions, plant trees, clean up rivers, and rewild land to mop up carbon. We must start funding farmers to safeguard our natural capital and provide public goods that include not only the food we expect, but also other things we value, such as nature, beauty, heritage and connection.

The government has recently introduced environmental land management schemes (ELMS) for landscape recovery, countryside stewardship and sustainable food production. But farmers point out that the numbers rarely add up and at times can be little more than greenwashing. Yet if done well, ELMS could help farmers rescue nature as well as feeding us.

Supporting positive work in this way complements punishing harmful practices. Policy needs to blend laws, subsidies and taxes. Ban the pollution of rivers. Subsidise hedgerow restoration. Tax unsustainable foods... This is all vital if healthy, sustainable food is to be affordable for everyone, while also tackling our commitments to reach net zero by 2050.

It is amazing to think that when we sit down for dinner this evening, we make personal choices related to the public health crisis, nature and the climate, as well as sending a message to supermarkets and politicians about what matters to us.

Somewhat overwhelmed by all these issues, I bought a local apple and left the farm shop.

My biggest hope when I saw today's grid square on the map had been the opportunity for a bracing winter dip in a lake created in a flooded quarry. I headed straight there from the farm shop but was disappointed to find the water barricaded by fences and barbed wire. Signs every few yards warned of 'Lake Safety', 'Danger Deep Water', 'No Swimming', and 'No Unauthorised Fishing'. You do need to be cautious when swimming in old quarries, but the fence that separated me from the natural world stopped me from making my own decisions.

I turned away, frustrated. Later research revealed I could have swum in the lake, in a neighbouring grid square, so long as I followed lots of rules, signed some paperwork, paid £30 for an induction course and then another £7.50 for my dip. That's not my kind of swimming, although the venue is a fantastic regeneration project and a sign of the growing appetite for people to swim, stand-up paddle (SUP) and canoe. This map was regularly making me consider the many sides involved in land access arguments.

I was searching for nature but kept being forced onto roads or sandwiched on narrow footpaths squeezed between high fences. I wanted to get into an area of woodland where old pits had scrubbed over since the quarry closed, but that too was cordoned off. I could only peek through the fence at a spot where someone had lobbed a TV and a raw chicken into the bushes.

So I turned my curiosity to the human world instead. Among the KFC cartons and Monster cans on the pavements was a discarded prescription box of tadalafil, a medicine used to treat erection problems. It cautioned, 'Check with your doctor before taking tadalafil if you have a curved penis.'

Sellotaped to lamp posts were adverts for a circus, a cat that had now been missing for eighteen months, a music event from last summer featuring a 'barn dance in a barn', a tattoo artist in the city, and the local branch of Slimming World ('Your slimming success starts here'). And I smiled at an angry note propped against a discarded dog poo bag, ranting, 'WHO LEFT THIS HERE IS AN IDIOT. THE BIN IS 50 YDS AWAY.'

I was close to conceding defeat on this grid square, where more than two-thirds of the open space was sealed off. I scrambled through a patch of overgrown thicket, bashing through birch saplings and nettles until I forced my way out of the chaos into a field of barley. The farmer had left a margin of fallow land around the edge, as even a metre-wide strip helps wildlife. As if to prove this, I came upon a fox deep in thought. I stood stock still and it was several seconds before the fox spotted me. Then it, too, froze, eyeballing me from twenty yards, as we dared each other to make the first move. My furry friend cracked first and dashed into the safety of the undergrowth.

Birds such as grey partridges, whitethroats, yellowhammers and corn buntings nest in these wild margins, as well as harvest mice and voles that provide food for kestrels and barn owls. These small hunting grounds away from roadsides also decrease the number of owls killed by cars.

I followed a sunken track up into some woods. I heard a squirrel chatter, a buzzard mewing and – as always in this shire – the thrum of a motorway. This morning I had made the foolish mistake of checking my emails before coming out. After having a quick play on a rope swing hanging from a high bough, I sat down to drink my flask of coffee and eat my apple beneath a majestic beech tree. But thinking about my emails overshadowed my appreciation of the crisp, peaceful day.

A message from my accountant had warned of a steady downturn in my fortunes. I was still earning enough, so my brain was not worrying about cash, but rather my dwindling motivation to hustle, to chase, to 'succeed'.

I've been doing the same sort of thing for so long now (I hesitate to call it 'work'): go on adventures, write about them, earn some money. But what would happen if I changed tack? What if I did something else? Should I keep going or change course? I had no idea.

My priorities had been evolving for a while, suggesting it was time to switch direction. But I was not sure which way to go, nor brave enough to find out. There was a hollow inside me that my expeditions used to fill, and I didn't know what would take their place, which was probably why I was sitting on my own in a wood with no whoomph to do anything beyond trying to enjoy this morning and this grid square.

I was languishing through this season of my life, wintering like the bare beech tree I was leaning against. Maybe it might teach me something if I sat long enough and listened?

I thought of the unusual Old English word *dustsceawung*, 'a consideration of the dust', or the contemplation of what we have lost and the transience of things. It reminded me not to fret too much about my email worries, and instead to savour this hot coffee, this cold morning, and this enormous tree reaching out above me and around me.

MUDLARKING

'Without willing it, I had gone from being ignorant of being ignorant to being aware of being aware.'

Maya Angelou

One motivation for exploring a square each week, come rain or shine, was to make being out in nature part of my routine. I hoped that becoming connected with where I live, with its weather and seasons, would keep me attuned to the seedlings pushing through pavements, the migrating birds passing overhead, the provenance of the food I eat, and reveal some interesting new running routes too.

Taking just a few minutes every month to climb a tree, which I'd done for the past three years, had certainly made me happier. Each time I returned to the tree I was surprised by how much nature had changed in the past few weeks. Fun, too, had been my year of full-moon forays, getting outdoors for a run, ride, walk or swim on every full moon, and also a year of enjoying coffee outside at least monthly. If hospital gardens help people to heal, if doctors now prescribe exercise in nature, and if the 'Natural Health Service' addresses a range of conditions, then committing to fifty-two outdoor missions sounded like a sensible undertaking. By now the habit of heading out once a week with my camera and notebook felt comfortably established.

It was a flat, grey day beneath a flat, grey December sky. The river flowing through today's square was flat and grey, rippling as the tide nurdled ever lower. My mood, however, was neither flat nor grey. I was looking forward to this one.

A few off-limit jetties jutted out into the current, jetties for pipelines and industry. A conveyor belt rumbled down one, filling a barge with gravel, but all else was quiet. This was, perhaps, a grid square that only a map nerd like me could derive pleasure from. More than half of it was blue on my map, but that was an incongruous representation of the muddy, intimidating industrial estuary spreading out before me. I didn't dare swim out to explore it.

Behind me, the rest of the square was fenced off by a shooting range, an electricity substation filled with fizzing power lines, a cement factory, a slime-covered canal (featuring a sofa tipped into the water, whose lurid colour perfectly matched the algae), and a police firearms training centre complete with replica streets, shops, houses, a pub, stadium and life-size sections of planes and trains. This brought back fond memories of getting a day's pay back when I was in the Territorial Army at university to don 'civvy' clothes and cheerfully lob half-bricks and milk bottles at massed ranks of policemen in riot gear. It was all fun and larks until they mounted their response charge at us…

And so, in terms of my exploration, the square was effectively reduced to little more than the footpath along the embankment's flood defences, plus whatever muddy 'beach' was revealed as the tide fell. That was fine by me as I'd studied the tide timetable and arrived a couple of hours before low tide, past a yard filled with ships' anchors, ten-feet tall and tonnes galore. I was here to go mudlarking among the slimy green rocks, brown seaweed and thick grey mud of the foreshore.

A mudlark is someone who scavenges in river mud at low tide, looking for valuable items. It was a way of life in London during the 18th and 19th centuries, when mudlarks searched the Thames' shores for anything of value. They earned little but enjoyed an unusual amount of independence for the period, plus they got to keep whatever they found or earned.

Lara Maiklem explores the ancient, murky, tidal foreshore of the Thames, whose ebbs and flows still churn objects to the surface that have been hidden and preserved in the mud for centuries. I had devoured her fabulous book *Mudlarking* (and enticing Instagram posts), and was fascinated by the greedy prospect of finding treasure, Roman roofing, Tudor shoes, and messages in bottles.

I donned wellies and waterproof trousers, climbed up and over the graffiti-covered embankment wall, and dropped onto the foreshore to begin my search. Its lowest reaches were a lethal gloop of deep, sloppy, stinking mud. I settled for making my way along the line where rock and mud meet, slipping over mounds of bladderwrack, a brown seaweed studded with air bladders that help it to float upright and absorb nutrients when submerged.

At low tides, the exposed seaweed forms dense beds, which theoretically should provide shelter for all sorts of creatures. But I'm afraid I saw not a single living thing among it all. A few gulls bobbed on the river, and semi-feral ponies grazed on the embankment behind me. But the water was pretty grim.

Only a few pearly-white oyster shells gave any suggestion of life in the grey mud. Over the past 200 years, habitat loss, pollution and overfishing slashed the oyster population around the UK by 95 percent, though it is now on the increase again. Across the country, things are improving from the low point of 1957, when the Thames was declared biologically dead and the river was a foul-smelling drain. It is

a travesty, however, that even today, not a single river in Britain is free from pollution.

I had fully intended to find priceless loot within minutes of beginning my mudlarking. Instead, I found a rusty chair frame and heaps of plastic, including a label saying 'BAG IT AND BIN IT, DON'T FLUSH IT'. I picked up a 1980s milk bottle with 'PLEASE RETURN BOTTLE' embossed on the glass. All interesting enough, but where was that jewel-encrusted sword when you needed it?

Truth be told, my patience began to wane within about twenty minutes, as I had known it would. This was actually one reason I'd decided to try mudlarking in the first place, to remind myself to slow down, to savour the process of searching, and not to be so hung up on productivity or getting things done.

So I persevered, picking my way among rusty pieces of metal, crisp packets and drinking straws. We used to throw away 4.7 billion plastic straws, 316 million plastic stirrers and 1.8 billion plastic-stemmed cotton buds each year. Those numbers plummeted once they were banned: proof of the immediate impact that quick, simple law changes can have.

I stood up straight to stretch my back and to watch a ship pass down the river, filled with the romanticism of imagining all the places for which it might be bound. Nineveh, perhaps? But my maritime musings have become more accurate, if less exotic, since I downloaded the Marine Radar app, which tells you about any ships you see.

So this was the Maltese cargo ship *Celestine* sliding down the estuary with a salt-caked smoke stack and a cargo of cars. Heading in the other direction, a Dutch trailing suction hopper dredger slurped up the same gloop I was searching through. Dredgers work like monstrous vacuum cleaners, sucking up sand, mud and gravel from the channel to store onboard and discharge later. I wondered what gems had unknowingly been dumped through its pipes.

I bent down again and kept looking. Now I found a metal fork, a white comb and the compulsory shopping trolley. How did they end up in the river?

A discarded condom, unopened, told its tale of a disappointed date lobbing it off a bridge on his unplanned lonely trudge home to an empty bed. A golf putter, green with slime, had me imagining a pitch and putt rage, a nice day out soured by a tantrum and the golf club arcing through the summer sky into the water.

What else did I find? A pair of red pebbles caught my eye. A smooth, tactile fragment of green bottle marked 'A.A. & Co'. Two symmetrical shards of tile. A fragment of porcelain decorated with blue and white lines, dots and circles.

That was about it.

But still, I was 99 percent certain that Christopher Columbus had dined off that very plate, munching corn on the cob as he set sail to discover Australia. One can always dream...

Even though I found no verifiable bullion or antiques, I had enjoyed trying to imagine stories for all the mundane objects I collected and brought home that morning. All these banal discoveries were grist to the mill as I learnt how to be an enthusiastic amateur. I was like the young boy Calvin in the comic strip, digging up the garden with Hobbes, his pet tiger. Hobbes asks Calvin what he has found.

'A few dirty rocks, a weird root, and some disgusting grubs,' answers Calvin from deep in his hole.

'On your first try?' asks Hobbes in delight.

'There's treasure everywhere,' exclaims Calvin.

LITTER

'This month taxes a walker's resources more than any... You can hardly screw up your courage to take a walk when all is thus tightly locked or frozen up and so little is to be seen in field or wood.'

Henry David Thoreau

December is a quiet month. I looked out over a rolling landscape of empty fields divided into squares by long, straight hedges. The only sounds were of distant cars. When did it become normal to hear more traffic than wildlife? Everything felt subdued. But I was glad to be beginning this project at the tail end of a year so that I could experience everything perking up in the brighter days to come.

With practice, a map becomes as clear as a picture, and almost as full of imagined detail. This week's grid square seemed to be as hilly as almost anywhere on my map, a rare treat for me in this flat, sanitised region. I adore land with contour lines, curvy and sensuous, filled with possibility.

A lone farm at the top corner was the square's only building. One minor road shaved the left-hand side. A hilltop ran along the eastern edge, along with 600 metres of bridleway, the only technically-permitted access on the grid square. I was looking forward to exploring all this empty space.

Even so, I found it hard to escape the tyranny of my mobile phone's frivolous temptations as I began walking.

'I'll just quickly check,' I thought. 'My emails...my social feeds... Leeds United...'

I had to force myself to put the phone away, chastising myself with the reminder that I was here to pay attention to the world, not to escape from it. That included acknowledging the ugly things too and trying to be interested in them. First up was a McCafé iced coffee cup (the latest addition to my McMenu of litter), a Coca-Cola bottle (the world's biggest plastic polluter most years), and a heap of fly-tipping.

Litter had been lobbed from cars in every passing place along the narrow road. I hated seeing litter on every grid square I visited. As well as the bottles and sweet wrappers, someone had tossed away a banana skin. That's far less of a problem than plastic, certainly, but fruit can still take a couple of years to decompose: the John Muir Trust removed 1,000 banana skins from the summit of Ben Nevis during a cleanup. This matters here because, in the heated arguments surrounding increased public access to the countryside, litter has taken on a totemic significance.

'Of course you shouldn't allow the filthy public to roam at will: think of all the litter they'll drop!'

'Of course you should allow the public to roam: only a few people drop litter, and that is because they're disconnected from the landscape. Educate them to care and everybody wins.'

And round it goes.

There are countless examples of public-spirited souls opening up their land, only to see it spoilt with burnt-out cars, piles of litter, torn-off branches and trampled bluebells. Increase the number of permissive access routes to create circular walks for people, and you risk off-roaders roaring around and cutting through sheep fences, turning peaceful and biodiverse woods into noisy, muddy trackways, and livestock being killed.

It is interesting (or interesting to me, at least) to consider the history of littering. After the First World War, American industry boomed. But this required consumers to buy more and more stuff to keep up with the pace of production and ensure profits stayed high. Since we only need a finite amount of junk, manufacturers began persuading us to throw things away and buy more. They started advertising, inventing new trends, building obsolescence into products, and encouraging a disposable culture. And so began our pandemic of littering.

A narrow focus on individuals who drop litter diverts attention and responsibility from the industries that continue to produce single-use products in vast quantities. America's 'Crying Indian' TV ad was incredibly successful at this in the 1970s. It's a similar trick to the tactic of BP, the fossil fuel company, which hired an advertising firm to help concoct the notion of the personal carbon footprint – and all the personal guilt that goes with it – to put the blame for climate change onto individuals and thereby distract from the true villain: the industry itself, with just 20 firms being responsible for a third of all global carbon emissions.

People litter for many reasons, including the social influence of whether others have dropped litter, feeling alienated, a lack of education, that it is easier to litter than not to do so, and a failure of enforcement. It is an ugly byproduct of the way we live and our increasing estrangement from nature and community.

Litter and disconnection were recurring issues as I explored my map, looping round and round in their own grubby version of a circular economy. Would tackling one stop the other? Are they keystone indicators of the way we mistreat the landscape, or are they red herrings that distract from more important matters?

Shaming litterers seems likely only to provoke more littering, rather than encouraging us to love where we live. Perhaps we need to positively encourage personal responsibility and to save our shaming for the profiteering manufacturers churning out plastic, for our throwaway society as a whole, and for our government for not enforcing change?

How can we build pride within communities, so that people won't tolerate living in litter? Those who believe that allowing access to the countryside will result in a pestilence of litter are quick to harrumph and shame the kids dropping drinks bottles, but rarely offer alternative ways for them to forge connections with nature and where they live.

The other side, who argue that access for all will be fine once the litterers are educated, also have no magic-bullet solutions to reach the masses.

Finally, how much does dropping litter matter in the grand scheme of things?

Putting your crisp bag in a bin rather than lobbing it into a field means only that it will be carted off to a landfill site or incinerator, often in a distant poor country. Its toxic impact on the planet is no less than if it lay in the field. We focus on the crisp bag but are blind to the overgrazed land stripped of wildlife.

A farmhand in his tractor cab didn't see me as he flailed a hedgerow to within an inch of its life. He was the only person I saw all afternoon. I waited for the tractor to turn away from me and then nipped through a strip of woodland into a field. I emerged at the top of a valley with a relatively huge expanse of open space before me. This was the closest I had come to 'proper' hills on my map so far. A joy and peace welled up inside me. I have a yearning for space and solitude that I underestimate until moments like this reminded me of what I had been missing this year.

A few crows flew across my eye line. Although we don't say a 'few' crows, of course, for we have the eccentric convention of animal collective nouns to enjoy. Thus, it was a 'murder' of crows that flew across my eye line. Most of these quirky 'terms of venery' date back to a 15th-century nun called Juliana Berners. She coined the terms in the first colour book in English, *The Boke of Saint Albans, Containing Treatises on Hawking, Hunting, and Cote Armour.* They include a wake of buzzards, a commotion of coots, an asylum of cuckoos, a curfew of curlews, a crown of kingfishers and a conspiracy of ravens.

But how was I to know if I was watching a murder rather than a conspiracy? How do you tell the difference between rooks and crows, other than knowing that crows are larger? The secret is to remember the old saying, 'a crow in a crowd is a rook, and a rook on its own is a crow'. The lonely 'caw' of a crow is synonymous in my mind with misty dusks. Rooks, by contrast, are sociable and noisy birds that nest in colonies and feed and flock together. In winter these groups can number hundreds. A mouthwatering sight, I imagine, back in the day when rook pie was a popular rural meal.

It was a dank winter's day, grey and two-dimensional. But from this viewpoint, the foggy haze felt cosy and private. Although I'd never been here before, the landscape felt familiar and reassuring. Green and pleasant fields are a classic pastoral snapshot and I grew up in countryside like this. In the years I spent overseas, my nostalgic memories were often of the traditional rolling hills of home. But much of this lush, fertilised land is actually starved of nature, planted with grass seed mixtures designed to maximise nutrition for cows and reduce feed costs, but at a cost in biodiversity.

After I read George Monbiot's *Feral*, with its bold claim that 'the sheep has caused more extensive environmental damage in this country than all the building that has ever taken place', I started looking at our countryside through a new lens, seeing a 'sheep-scraped misery' and a sad emptiness to those upland hills I adored. The late ecologist Frank Fraser Darling labelled our uplands as 'wet deserts', such is the lack of nature on the overstocked, overgrazed, damaged land.*

But I am so accustomed to neat green countryside that when I first visited the Knepp estate in Sussex – the pin-up trophy of the British rewilding movement – I was disappointed by how scruffy it looked. There's a fabulous one-star rating on TripAdvisor saying that Knepp 'resembled a piece of wasteland such as you would find behind an industrial estate'.

We have ingrained ideas about the countryside and farming, influenced by our shifting baselines and dating back to the picture books we enjoyed as children with happy farmers and their happy animals. But the truth is that many modern farms are factories not fairy tales.

Although these fields were grazing pasture, I couldn't see any cattle. They had been moved into winter sheds to avoid damaging the sodden ground and to be fed a high-protein diet that would boost milk production towards anything as high as sixty litres per day. As farms have grown, it has become easier to take the grass to the cows than the other way around.

When I walked past the farm later, I had to wade through slurry that oozed under the cowshed doors. Keeping animals indoors, as opposed to grazing them outside all year, is time consuming and also entails high fossil-fuel costs to make silage, feed the animals, and then remove their slurry, which is a pollution hazard.

**Pockets of the Elan Valley in the Cambrian mountains are known as the 'Desert of Wales' because of the dominance of Molinia grass and the lack of biodiversity caused by overgrazing and clearing.*

I did eventually see some cows near the end of my walk through endless fields of rye grass. About twenty ran towards me on the footpath to say hello. I'm fond of cows and like having a chat with them, but I erred on the side of caution as they galloped my way. I hurdled a fence with a yellow sign 'Warning Bull in Field', and said hello from behind the fence, preferring to take my chance with the lone, uninterested bull rather than the flighty females. They came to a halt in front of me, chewed the cud lugubriously and peered at me with big brown eyes.

Apart from that minor excitement, the rest of the walk was pleasant but uneventful. I hiked down a hill, up the other side, down another hill, and then back up again. I wasn't used to such exertion. I enjoyed it though, my muscles warming with the effort and my eyes taking in the ever-changing views.

I stopped for lunch at the highest point of the square, sheltering under a scrubby hawthorn tree to keep off some of the drizzle. Millions of water droplets, tiny shining globes, gleamed on the blades of grass. I pulled the Thermos flask from my rucksack and poured a steaming mug of bright red, homemade beetroot soup. It was hot and colourful and the perfect antidote to the cold day's monochrome weather.

Where I live, there are an average of four hundred people squashed into every square kilometre, a hundred times more than in Iceland. But here was I, on the quietest of hilltops, with almost a whole grid square to myself, slurping soup. I felt lucky to have got off the busy and beaten track today. All these hills and fields disappearing into the mist had even been sufficiently beguiling to help me resist the habit of reaching for my phone to 'just quickly check...'.

CREEKS

'When we try to pick out anything by itself, we find it hitched to everything else in the universe.'

John Muir

Litter was strewn over today's grid square like wrapping paper on Christmas morning. I didn't want to be disheartened by it on every outing, but nor did I want to *not* see these problems or accept them as normal and just shrug my shoulders. I live far from the cascades of contour lines, miles of moorland or rushing rivers that I relish. Could I really scratch my adventurous itch on this tame map, bereft of mountains, oceans or dragons? I doubted it this morning and wished I was exploring Siberia rather than suburbia and this odd ecotone, a transition area between the city, suburbs and countryside.

Out of all the country's maps, mine was down in the rubbish relegation zone for adventure potential. But that also made this project a more universal one than if I lived, say, on map 402 in the scenic Scottish Highlands or map 24 in the Peak District.

I would have preferred to try this experiment in lots of other places. But I suspect it is true for many of us, that a perceived overfamiliarity with where we live leads to either contempt or, at least, 'unseeing'. We suffer from place blindness when we spend a lot of time in familiar environments without being observant.

What might happen instead, I asked myself, if I worked on what Mary Oliver described in her poem 'Going to Walden' as the 'slow and difficult trick of living', and finding it where I was? To cherish the changing seasons here as I did when I was cycling across continents, and to relish the rough-and-ready backstreets and everyday life as I always did in foreign lands. Perhaps it might even help me to manifest a sense of belonging that I had been missing for such a long time.

Lying on top of a loaded skip was a four-foot painting of a waterfall, forest and mountains, in a shiny gold-painted frame. When a leaky roof caused a homeowner in Toulouse to repair his attic in 2014, he discovered a lost painting by Caravaggio, valued at up to £100 million. Could this skip painting be my winning lottery ticket, an old masterpiece tossed out with the trash? Granted, the painting was not an easy picture to admire, and the frame was particularly tacky. But I resolved to take my discovery home with me at the end of today's stroll, and then retire to Monte Carlo.

The estate of modern homes and flats where I found the painting stood on the site of one of England's first paper mills. Apart from the name of Paper Mill Lane and an artificially straightened river, there was no trace of the bustling riverside wharf that thrived here for centuries until it closed with the loss of more than a hundred jobs and many more stories. In its place today were these residential cul-de-sacs and an enormous supermarket transit depot.

I learnt about the paper mill from an old chap who saw me photographing the skip, said hello, and then poured forth a stream of proud facts about his neighbourhood, like a territorial robin singing on the garden fence.

That the river here used to be sixty feet wide, and busy with boats.

That Elizabethan bricks were tiny.

That the historical site had been filled with foundation cement overnight by developers pre-empting archaeologists protesting the proposed residential development.

That the street down there (he gestured to a row of smart old semis) had been for the mill managers.

That the wrought-iron fences ringing the street's trees were removed in the war to be turned into munitions.

That he and his mates used to run across the road over there for dares (now a busy dual carriageway), until one of his mates misjudged it a bit and got killed.

That another of his classmates robbed an abattoir, and the police shot him dead. And that he then joined the Marines to escape all this. ('I arrived with my hair down to here; they gave me a number one buzzcut all over!')

That he climbed ropes, ran assault courses, and came home to his mum on leave with knots of muscles and a new sense of hope for his life. 'It sorted me out, the Marines did. I joined the Fire Brigade after that until I retired. But there's nothing for the kids to do here anymore. No youth clubs, no sports fields, even the churches are locked six days a week. Anyway, good luck to you, son. There's so much history for you to find round here. Merry Christmas too.'

This week marked the midwinter solstice when the sun is at its lowest point in the sky and we have our shortest days and longest nights. Many modern Christmas traditions have their origins in the ancient midwinter celebrations. It has always been a time to hunker down and huddle round fires with our loved ones.

I caught a whiff of apple *shisha* from an open window. The sweet smell of the smoke whisked me away from Christmas to Ramadan in Beirut, memories of warm evenings strolling under palm trees along the seaside corniche. A long way, then, from the empty bowling club I was now peering at over a high fence. The neat green lawn held its own gentle appeal, though, and I looked forward to being old enough to justify taking up the game.

Opposite a shop advertising 'The Ultimate in Mens Outsize Clothing', I cut down a ginnel through some industrial units, past a sewage works, and out onto flat fields by the dual carriageway and the distribution warehouses. It was a damaged and abandoned land, artificial and empty. There was nothing here. No nature, but no man-made purpose either. Soggy grass stretched thinly towards the riverbank, patterned by the skids and swerves of dirt-bike tyres. Crushed cider cans. A McDonald's Sweet Curry Dip (whose ingredients began 'Water, Glucose-Fructose Syrup, Apricot Puree Concentrate, Sugar, Spirit

Vinegar, Modified Maize Starch'). Artificial and empty.

I followed a large metal pipe from the water treatment works to its discharge point in the creek. Nothing felt very Christmassy today. A fence was wrapped, not in tinsel, but in blue-and-white police tape that flapped in the chilly wind. 'POLICE LINE DO NOT CROSS'. Graffiti on a concrete barrier urged 'Fuck Boris Big Up NHS'. Another cautioned 'You aware Covid Bill Gates hoax +5G'. Even allowing for the many tenuous segues and questionable meanderings of these chapters, I couldn't bring myself to dive down the online depths of that zany conspiracy theory.

The river flowed slowly under a bridge beneath the dual carriageway, and I followed it onto the marshland beyond the road. It felt as though I had crossed a boundary, passing from the scraggy edgeland out into a wilder, forgotten world. The river was about ten metres wide and flowed straight through the flat marsh beneath strings of pylons and a clear blue sky. At high tide, it looked calm. By the end of my walk the tide was falling fast, the water was murky and churning, and the banks were sheer, slippery, hazardous mud.

This scrap of freshwater marsh, surrounded by industry, was a complex mixture of wet grassland, ditches and scrub. It was a haven for breeding and overwintering birds. Marsh harriers, bearded tits and warblers abound in the summer. There were rare wetland plants and an important population of the threatened water vole.

Today I saw flocks of wading birds – redshanks, lapwings, dunlins and oystercatchers. Greylag geese and shelducks grazed on the grass, while teals, wigeons and cormorants dried their wings in the sunshine. Clouds of terns rose as one as I walked past a flooded industrial excavation. They swirled in the breeze and then settled again on the water once they deemed me safe.

I have grown to love birds in recent years, but still haven't got very interested in ducks, geese or gulls, even after relishing Adam Nicolson's book *The Seabird's Cry*, about these 'wind-runners and wind-dancers'. I'm not sure why. My unfair dismissal of 'seagulls' may come from associations with birds pinching my chips at the beach with scornful indifference, and so seeming more like pests than free-living, freewheeling 'wind-spirits'.

That prejudice was reinforced here when I turned left and followed a small creek up to the council's 'waste reception centre and transfer station'. The tip, in plain English. I watched through a security fence as men in big diggers shunted heaps of stinking bin bags around and herring gulls squabbled over nappies and pizza boxes. One gull was trapped in the narrow space between two wire fences, unable to take off vertically enough to escape. I watched the bird flap in distress but could not reach it to help.

A heron took flight from the bays of rotting rubbish, also preferring the tip's easy pickings to the effort of organic hunting in the creek. The tempting distraction of cheap calories. The bird bent its spindly knees, flapped those six-foot grey wings, and launched itself skywards to circle away over the marsh. Seeing that huge, prehistoric-looking bird scarfing junk food on the tip dismayed me. It seemed undignified somehow.

A sturdy wooden barge with tall wooden masts was moored a little farther up the creek. I googled her name and learnt she was almost 150 years old. I wondered how she had ended up moored here next to the roar and stink of diggers and landfill. These sailing barges were common sights on the river for 500 years, being the largest vessel that just two men could handle. Their shallow draught made them ideal for nosing up narrow tributaries and creeks to the mills and small factories.

I scampered across the dual carriageway back into residential streets

whose windows shone with Christmas trees and fairy lights encouraging the return of the sun. A sign taped to a fence offered a reward for a 'missing grey parrot (microchipped)'. A billboard advertised a church ministry where 'young and seasoned professionals are groomed to become urban missionaries in their careers and businesses. They display unique dimensions of divine excellency in the very heart of the marketplace.' A small, handwritten sign directed 'funeral flowers this way'. And a naked plastic doll sat with outstretched arms on the roof of a garden shed.

The doll stared down at me as I completed my circuit of the grid square, lifted the abandoned painting out of the skip and headed home for a mince pie or two, the daylight already leaking from the cold midwinter sky as the calendar year faded towards its close.

JANUARY

GARDENS

'Each new year is a surprise to us. We find that we had virtually forgotten the note of each bird, and when we hear it again it is remembered like a dream, reminding us of a previous state of existence. How happens it that the associations it awakens are always pleasing, never saddening; reminiscences of our sanest hours? The voice of nature is always encouraging.'

Henry David Thoreau

The darkest hour may be just before the dawn, but the darkest morning comes well after midwinter, when the jollity of Christmas has long since faded away. The latest sunrise is almost three weeks after the December solstice. It might be a fresh calendar year and a new start, but as I cycled out today it was one of the bleakest weeks of the year, with barely eight hours of daylight on my map.

The January sun, when it eventually showed up, skulked low and reluctant across the sky. There had been a roaring in the wind all night and the rain fell in floods. And now in the morning I was on my way masochistically to what looked to be one of the most nature-depleted squares on my map, in one of the most nature-depleted countries on the planet. This crowded map lies on the outskirts of a large city, so there are many pressures on its space, including farming, transport, industry, housing, and recreation. Everywhere you look, you see human impacts on the landscape, ranging from landfill sites to relaid hedges.

There was little need for the cartographer to use any green ink here; the whole square was a grey grid of boxes representing buildings. Colour came only from two busy roads, marked in yellow. There were just four scraps of footpath, little more than a couple of hundred metres of cracked tarmac, broken glass and dog mess. I felt in more need than usual of nature's gladness, but could I find any of it here?

The tragedy of the commons, that individuals ignore what is best for society in pursuing personal gain, suggests that humans cannot manage a common resource. Why do we care so little about the earth? Is it because we assume it is limitless? Apollo astronaut Edgar Mitchell's perspective changed after flying to space. He said, 'You develop an instant global consciousness, a people orientation, an intense dissatisfaction with the state of the world, and a compulsion to do something about it. From out there on the Moon, international politics look so petty. You want to grab a politician by the scruff of the neck and drag him a quarter of a million miles out and say, "Look at that, you son of a bitch."'

Why do we care so little about nature and its tragic decline? Is it because we have stopped noticing it? It is not that we have a short time to live, but that we waste a lot of it. It is not that the world is too small, but that we miss so much of it.

As I rode up and down today's terraced streets, I passed a stout tree sawn off at thigh height. A dog, pressed up against a window, barked incessantly, desperate for fresh air and bigger horizons.

Many houses had their gardens concreted over, prioritising ease of maintenance and parking spaces. Others had replaced front lawns with artificial grass, which looks tidy and is easy for busy people to deal with. It has become a vast industry, worth a whopping £2 billion per year. But the green plastic is not very green, for it blocks access to the soil for insects, starves creatures living in the soil, and provides no benefits for nature. It is another pressure adding to the precarious state of our insects, with more than 40 percent of species declining and a third endangered.

Our government is already missing the targets it set to protect nature, and not many of us seem particularly bothered. Although we like the idea of a green lawn, a lot of people don't seem to mind whether it is grass or plastic. But children can't make daisy chains on plastic lawns.

Fortunately, as my irritation grew about the mayhem we're wreaking on the world, my eye was drawn by an evergreen hedge, bursting with yellow and orange berries: firethorn. The owner saw me photographing it. 'The birds love it too,' he called from his driveway. 'It's fantastic. I planted it to stop the kids sitting along my wall, like birds on a wire. Shouting, dropping sweet wrappers, all that carry-on. Birds don't do that. It's been great, has that hedge.'

His firethorn hedge then opened my eyes to other gardens as I pedalled up and down. A number of people had made concerted efforts to remove all traces of the natural world from their properties, only to hang baskets of fake plastic flowers around their doors. But many gardens did have grass, plants and bushes in them. Now that I was concentrating, it was apparent what a significant area they all added up to.

Together, Britain's gardens are larger than all our nature reserves combined. These tiny oases are vital havens for wildlife amid the concrete jungle of our cities and the sterility of our farmland. If we have a garden, we can make it a little wilder and help to provide corridors for animals to move along. That is why Wild East, a regional nature recovery alliance, has put together a 'Map of Dreams' showing everyone in the area who has pledged to rewild their garden, churchyard, school grounds, farm or business and so generate more space for wildlife.

Nudged to notice, I kept searching for nature throughout my ride. I found noisy starlings, black-headed gulls, and a fox's path pushing beneath a fence. There were clusters of the weed called annual mercury, also known as girl's mercury or boy's mercury because, according to Pliny (our regular correspondent), pregnant women could use it to help select the sex of their child.

Ants and bullfinches enjoy the seeds, and in Germany some people boil the leaves to eat, for its acrid taste dissipates as it cooks. But in France they call the plant *la mercuriale annuelle* or *la foirolle*, and '*la foire*' means diarrhoea. A German's delicacy, perhaps, is a Frenchman's laxative. The choice is yours.

As the T-shirt says, you can't buy happiness but you can buy a bike, and that's pretty close. It was a pity then that I didn't see any kids riding bikes around town, and the only adult cyclists were the fast-food couriers with enormous delivery bags on their backs.

One of the town's numerous boarded-up shops was a failed bike shop, down the road from a closed-down public toilet that was disappearing beneath a shaggy mane of ivy cascading from the roof. Urban rewilding warms my heart (unless I need the loo). Rewilding is important in urban environments, and the United Nations declares that 'green spaces need to be placed at the heart of urban planning'.

Pleasing, also, was an ornate water trough and drinking fountain, carved from granite in 1903 in memory of someone's husband 'for the benefit of horses and dogs'. Today it is marooned by traffic on a roundabout island, surrounded by a world unimaginably faster and noisier than the one it was built to serve.

As I pedalled around, I mused that one of the contributing problems to the issues I'd encountered today was traffic. If you could magically get rid of it, there'd be a lot more space in the town for nature, and for children to ride bikes. Most houses had two cars parked outside, and every road was lined with cars. Removing cars might sound radical and unsympathetic to people's needs, but our car dependency is a relatively new phenomenon and our assumption that it is unavoidable is another shifting baseline.

When that water fountain was built, these streets were not jammed with cars, and the gardens were more likely to be full of vegetables than plastic grass. Households with two vehicles use each car for less than 5 percent of the time. Imagine if a self-driving electric vehicle could drop you off at work, take your family somewhere else, and then pick you up again in the evening to go to the pub. No family would need more than one car. Share that vehicle with your neighbours, and both the cost of commuting and the number of cars drops further. Invest heavily in public transport and even fewer cars are required.

So I felt oddly hopeful as I considered whether this might just be a brief blip of a few decades where every adult needs a car. Could it fade away into a new era of fewer, self-driving, electric vehicles with more of us making our local journeys by bike, foot, and electric public transport? We could then increase green space and community facilities.

Hungry now, I looked for a place to buy a snack. Someone had put out an old toaster on their garden wall, offering it to anyone who wanted to take it. But I had neither bread nor plug socket, so I rode on to a parade of shops. The car park was congested and the bike racks were empty, despite the stores targeting locals. Food has a direct impact on two subjects I'm interested in – the environment and our physical fitness – so I was interested to see what was available in this grid square.

The parade of shops was well placed to serve the local community. But I found only a post office, a rubbish convenience store full of biscuits and crisps, three takeaways, a beauty studio and a hairdresser. Britain has an appalling approach to community and food, with all the consequent impacts on our public health. We have become an overweight, underactive nation, living on processed calories and detached from both the land on which food is grown and the wide-ranging implications of our dietary choices.

This grid square was a fresh-food desert where access to fresh, affordable food was limited by inadequate public transport and unhealthy shops. While they are deserts, such locations – where more than a million of us live – are often simultaneously 'food swamps', obesogenic environments with high numbers of takeaways and empty calories.

More than half of our food these days is processed – the most in Europe – as sugar, fat and salt are hard to resist and cheaper per calorie than healthier foods. The price is an important factor for communities struggling with a cost-of-living crisis. It is much harder to eat healthily when you are poor, and rows of convenience stores and takeaways don't help. By 2050, the health costs of obesity are set to approach £50 billion per year. Residents in this grid square have to drive or catch a bus to visit a big supermarket, which is difficult for people on low incomes who are short on time and less likely to have a car.

It is crazy that our society is being harmed not only by too many cars and too many calories, but also by rising poverty that tends to mirror food deserts and swamps. Up to 15 percent of British families can't always obtain an adequate quality or quantity of food in socially acceptable ways, and more people than ever are turning to food banks to deal with food insecurity.

There was a lot to take in on this small square. I had begun the day with my mood matching the winter gloom. I felt sad about our lack of nature, crowded car culture and the impact of our terrible food system. But I'd also realised that, even in the depths of winter, there were plants everywhere, ready to let rip if we just left them alone. Insects and birds would soon follow.

It was not an affluent area, and the gardens were small. But nature was still there, willing to take any chance it was given. My favourite gardens had been those that crammed in a tree or a few bushes, leaving enough space for a couple of chairs and a summer barbecue. Gardens can brim with life, require little maintenance, and cost less than plastic grass or tarmac, as well as being good for the soul.

STILLNESS

'Look, and look again.
This world is not just a little thrill for the eyes…
You have a life – just imagine that!
You have this day, and maybe another, and maybe
still another.'

Mary Oliver

Busy days and rain falling. Chasing my tail and going nowhere. Horizons closing in. Boring routines and putting away the weekly shop. So, when I got an opportunity to escape, I bolted for the woods that the random-number generator ordered me towards. I bustled around grabbing my camera, rain gear and Thermos flask. I sloshed a can of tomato soup into a pan, added a tin of chickpeas, a handful of frozen peas and sweetcorn, and a glug of chilli sauce. By the time I was swaddled in waterproofs, my improvised lunch was hot, and I was out of the door and onto my bike with a smile returning to my face.

I ducked out of the rain into a beech wood. The trunks were slick and black, and the dark mesh of silhouetted branches was stark and tangled against the grey sky. Dead leaves had blown into heaps around the old parish boundary marker stone. But for the first time, there were also signs of rebirth.

Soft green moss squelched as I walked over it, one of 20,000 species that have been absorbing moisture, colonising new land, and regulating temperatures for 450 million years. Primrose shoots peeked out of the ground, the name of the flower deriving from the Latin *prima rosa*, meaning the first rose of the season. The world was beginning to wake from its hibernation.

I'd been out planting trees over the weekend with Trees for Cities, and my shoulders still nursed a satisfying ache from the work. Walking through woods on a sunlit summer morning is a pleasant experience, a simple hedonic enjoyment. But grafting away in winter to plant dozens of trees that eventually will become a wood worth walking through generates what is called a 'eudaemonic happiness', as it is also filled with meaning and purpose. Combining hedonic and eudaemonic happiness has the maximum impact on your levels of nature connectedness as well as making a positive impact on society.

When I entered the woods, I had passed various threatening signs hammered to trees, shotgun-pocked and rusty. NO TRESPASSING, DANGER SHOOTING, KEEP OUT. I was still in the early stages of this project, but already the amount of land I was forbidden from exploring was riling me.

Compared with most people, I suspect I have a relaxed attitude to access laws. I grew up in a village in the Yorkshire Dales and spent my childhood roaming the fields and woods and rivers with my brother and our two best friends, who were farmers' sons. We didn't leave litter or start fires. We didn't knock down walls or harass livestock. Of course we didn't. But we did explore absolutely everywhere. And nobody minded.

I am looking at the Ordnance Survey map of that village as I write these words, and it seems strange to see it all laid out formally, with just a few dotted green lines of public footpaths allocating the limits of where my childhood should have been technically allowed to roam. I would know that landscape so much less and care for it less (and have missed out on all that fun) had we heeded those restrictions.

From Yorkshire's green hills, I moved on to university in the emptier landscapes of Scotland. There were official footpaths there too, and I was thankful for them on the difficult, ankle-twisting terrain of Rannoch Moor or the steep valleys leading to the mountains around

Torridon. The difference is that those paths helped me to access the places I wanted to explore, rather than restraining me from all the places I was not allowed to go. Scotland has much wider access rights than England, though you must exercise them sensibly and not do daft stuff like walking through crops or gardens.

While Scotland enjoys an Outdoor Access Code that gives people a right to roam responsibly, 92 percent of England's and Wales's landscape is out of bounds for most of its population, and not many rivers have an uncontested right of access for swimmers. Almost everywhere in the country, you are a trespasser risking expulsion if you amble through a wood or paddle in a stream.

It strikes me as ridiculous that once upon a time somebody declared, 'This wood is now mine. This lake and this hillside too. I am claiming it. You may not come here anymore. Get off my land.'

Nobody owned woods or lakes or hills until the day they were first claimed by someone richer or more powerful than their neighbour. There is an injustice and an absurdity to being excluded from our wild places.

We have an excellent network of public footpaths, but they are mere threads across the canvas of our country. A public right of way is literally just that: a right to make one's way from A to B, but not to do anything else along the way, and heaven forbid that that should include lying down and sleeping! As I write this, a millionaire hedge fund manager who owns a 4,000-acre estate on Dartmoor (to accompany his 16,000 acres in Scotland) is challenging the wild camping provisions in the Dartmoor Commons Act, the last vestige of land in all of England and Wales where wild camping is officially tolerated.

Scotland filled me with a passion for the countryside and an understanding of my responsibilities to tread lightly. This then sent me off onto adventures in the expansive freedom of lands where you can pretty much go where you like, so long as you're able to take care of yourself and move safely through the environment: Siberia, Yukon, Alaska, Oman, Patagonia, even the frozen surface of the Arctic Ocean. These are places where you can feel part of a landscape, at best, but certainly never a master.

It is perhaps not a surprise that I have subsequently found living in the crowded southeast of England to be claustrophobic. I find it jarring to live in a culture that does not connect with the land in ways I have always taken for granted. This has played a large part in me making few friends and having little sense of belonging. Nobody I know here would consider running across miles of open ground for the fun of it, swim in a river at sunrise, camp on a hill or make coffee in a wood. Barely anyone, except dog walkers, even uses the footpaths, let alone laments how limited their access is.

And yet, despite the dearth of tree-huggers or cross-country runners, I have never seen so many intimidating 'Keep Out' signs festooned across a landscape, nor so many piles of dumped rubbish. When people don't feel responsibility towards the land, I'm not surprised a few are inclined to drop litter and leave gates open. If landowners are restricting access in order to keep their land clean and pleasant, it's not working.

As I tried this year to survey my map more thoroughly than just following the official paths, I was surprised by how much the issue of access impacted on the experience. How could I learn to love this landscape if I wasn't allowed within it? How would I be motivated to care for the natural world if I did not feel part of it? I had always assumed roaming across the countryside was an inherently normal thing, like breathing, until I began to get to know this map.

I had caused no one inconvenience or harm today. And yet I was trespassing, and I didn't like how that made me feel. I didn't believe I was doing anything wrong, but I have been conditioned to believe that trespassing is naughty and therefore I should feel guilt, not pleasure at being here.

Signs threatening that 'trespassers will be prosecuted' are mostly meaningless. Trespass is a civil rather than a criminal matter, so long as you don't damage any property. I had decided over this year to do my best not to be flustered by aggressive signs, to explore with consideration and care, but also a degree of freedom. If anyone got angry with me, I'd try to have a polite chat about it and then heed the perennial wisdom of *Three Men in a Boat*, written way back in 1899:

'The proper course to pursue is to offer your name and address, and leave the owner, if he really has anything to do with the matter, to summon you, and prove what damage you have done to his land by sitting down on a bit of it.'

And so, whether or not I was in these woods illegally, I continued up the path. I knew that if anyone challenged me, I would most likely be able to smile, chat and bumble my way out of any problems anyway. I knew also that I might be treated differently if I was a person of colour or a woman. In that sense, the outdoors is still not equally accessible to all of us, whether commoner, landowner, or of other ethnic or socio-economic groups.

I had been voyaging around my map for ten weeks now, which meant I'd covered just ten of its 400 grid squares, a paltry 2.5 percent. The vastness of my small map was becoming ever more apparent. Each week I discovered unknown places in what I had considered to be a largely familiar landscape. I would wager I know this map's paths better than most, and yet it was clear now how little I really knew and that I was seeing the place almost as if for the very first time. It made me appreciate the ancient network of paths that helped to guide me over the area.

Almost every footpath on my map has been there for centuries. Some have been walked for an astonishing 7,000 years since their origins as connections between Bronze Age encampments. These ancient green ways tend to follow the contours of the landscape. Then came the Roman routes, striding in efficient straight lines across England, moving troops and connecting infrastructure. After that were the Anglo-Saxons, settling down, working the land and establishing many of our current towns and villages. It feels precious to walk in those long chains of footprints.*

Our footpaths have a magnificent history. Yet 49,000 miles of them are due to disappear in England and Wales unless action is taken. The government has set a cutoff date of 2031, after which any unclaimed rights of way will disappear for ever. Reinstating them would boost our path network by a third. The Ramblers Association urgently needs online volunteers to help to apply for these thousands of miles of paths to be restored, protecting them for generations to come.

I left the wood and pedalled past a pair of pale-blue shepherds' huts

**A new movement, Slow Ways, is creating a national network of walking routes connecting thousands of Britain's towns, cities and villages. There are already more than 8,000 Slow Ways stretching for more than 120,000km, created by online volunteers. www.slowways.org*

to a compact 14th-century church and churchyard. Gravestones stir my imagination and I enjoy meandering among them. I calculate people's ages, feel sad about those who died young, look for my birthday among the dates, pause at those who died around my age, and think about all those couples 'reunited at last'.

Today there was '"Mick", Micheal [sic] Miller, Lost Not Forgot', with a statue of a fat, smiling Budai in this Christian graveyard. Budai is often called 'the Laughing Buddha', but he was actually a 10th-century Chinese monk beloved of children and the poor for giving out sweets and snacks from his sack. There were well-tended plots with little white railings, heaps of flowers, and photos of Mum and Gran. There were Bill and his 'old sweetheart' Jean. Danny's gravestone celebrated that he 'enriched our lives with his courage, humour, goodness and love. A real nice guy.' A tragedy, then, that he died at the age of just 31. Another epitaph read, 'He that wins knows no quitting.'

A rain-soaked envelope was addressed 'To My Darling Neil', the writing smudged like teary mascara. I imagined the contents of the letter, perhaps things left too long unsaid, a broken heart, regrets, apologies or consoling memories.

A fresh grave was mounded high with dark earth and covered with red roses and white lilies. A note lay on the flowers. 'Thank you for 61 happy years together. Goodbye my love.'

Even in a graveyard, our stories don't live for ever. They are places of immanence but not permanence. I found upright headstones, flat or kerbed ones, and table tombs. They were polished, part-polished, honed or pitched. But farther down the rows they were faded, moss covered, and then fallen over, with each story fading away and eventually forgotten. Older tombstones had been ensnared by green fingers of ivy* that climbed and entwined and reclaimed, as the wild world marches remorselessly over our fleeting presence.

I wondered why, if I like graveyards, do I hope my ashes are one day sprinkled into a swift stream or tossed into waves on a sunny morning,

**Ivy is clingy, luscious and misunderstood, with an exaggerated reputation for strangling trees and cracking buildings. Ivy has its own root system, so isn't parasitic, and its evergreen, woody tangle supports around fifty species of wildlife. It used to be said that a wreath of the plant would stop you getting drunk, hence the god Bacchus wore his wreath of ivy and grapevines.*

and that will be the end of it, and everyone will go home with a smile to get on with their lives. In the words of American writer Brian Doyle, 'these memories do not make me sad or nostalgic but rather thrilled and happy that I had those hours. No man ever savoured those hours in the game more than I did, no man in the history of the world; and rather than sigh at their loss, I sing at their gain.'

Birdsong is easier to single out in winter than at noisier times of the year. All I had heard today were the usual robins boisterously defending their patch, blackbirds' indignant flurries and wood pigeons repeating over and over their plaintive cry of 'my toe hurts, Betty'. Then a sudden burst of birdsong in a hedge caught my attention. It was a raucous outburst of disagreement between half a dozen mistle thrushes, according to the Merlin app, a genius tool for identifying birds by their sound. Sure enough, the chunky, pale birds bounded out of the hedgerow, across a soggy ploughed field, and then flew to perch at the top of tall trees and continue their fluty song.

From looking at my map beforehand, I'd expected that today would be a free roam through woods and empty fields. But most of the land belonged to a country house and had been fenced off for horse livery. Public footpaths feel a cop out when they're tightly fenced, as though the letter of the law has been heeded, but very much not the spirit. Rather than helping me to enjoy the countryside, these paths were just funnelling me to go somewhere, anywhere, but not here. Move along please, there's nothing to see here.

I sat on a log in a wood to drink my soup. I thought about how I'd rushed to get here today, how dissonant it was to be hurrying and clamouring to get into the stillness and calm of the countryside. Forcing myself to slow down helped me to resist trying to 'make an adventure out of something whose most important meaning is altogether more intimate and homely', as Richard Mabey put it in *The Unofficial Countryside*. This was at odds with the way I've pursued most journeys in my life.

Mabey's pursuit of that intimate and homely sense, which I interpreted as a sense of belonging and connection, was more likely to be caught 'in lunch hour strolls, weeds found in a garden corner, a bird glimpsed through a bus window. It was a change in focus that was needed, a new perspective on the everyday.'

In *The Runner*, Markus Torgeby wrote, 'I must do something about my restlessness. One day I put on several layers of clothes, sit down on a tree stump and do nothing. I must get over this hurdle, I must learn how to do nothing.'

I decided to give it a try myself today, here in my grid square.

That time on the stump, Torgeby concluded, 'was a good investment. Life became greater after that. Food tasted better and the song of the birds in the woods was even lovelier.'

Spending time alone in nature is a staple part of Outward Bound training for young people in the United States. The extended 'wilderness solo' experience, as it is known, derives from ancient traditions. Moses, Jesus and Buddha, as well as Gandhi, John Muir, Thoreau and Alexander Supertramp, all went into the wild in search of understanding and transformation. It has been an element of many societies' initiation rituals and helps them to clarify their strengths and purpose.

Today, aspects of these experiences have become part of wilderness therapy and outdoor education. Silence plays an important role in our development, and wilderness solo experiences can help counter the loneliness, stress and depression of modern lifestyles.

I spend a lot of time alone, but almost none of it sitting still without distractions. So I would start today with just one hour. Attempting even that felt somewhat daunting as I switched my phone to airplane mode and set an alarm to rescue me in sixty minutes (if I stuck it out that long, which I doubted I would). I sat down on a log and waited.

With pen and paper in hand, I would gladly sip coffee and sit on

a stump all day. Give me a book to read and it would be my dream holiday. But with no way to record my thoughts, I was left with the gentle maelstrom of nonsense inside my head for company. There was no escape.

One hour on a wet log included me racking up a dozen practical plans and a handful of emotional mood swings, while every minute involved observing something new in the wood around me.

I was bored. I fidgeted. Nothing happened. Minutes dragged.

I scratched my back. I thought about giving up. I needed to pee.

My brain was in overdrive, yet I also dozed off for brief moments.

I pondered how odd it was that I think the world of woodland and birdsong, but now that I was here, I wanted to go home and Tweet about it instead.

Most animal species have about a billion heartbeats in an average lifespan. It is up to me whether I expend mine feeling uptight about things beyond my control, or by savouring the everyday nature around me and doing what I can to change what I can.

Every hour is a substantial chunk of time and too precious to waste. The summer shriek of swifts will one day go out of my hearing range. I will put down my child one day and never pick them up again. I must not waste my minutes.

To my surprise, when the alarm eventually sounded, my first

emotion was of disappointment rather than relief. Sitting on a log for sixty minutes had been a surprisingly cathartic experience. From feeling flustered about the shortness of time and how little I was achieving, it was reassuring to have watched an hour flow out before me like a river, and to have chosen to be OK with that.

Settling in for a second hour would have been far easier. I felt relaxed and buoyed, and resolved to return to my sit spot and try this again soon.

As the monk Thomas Merton reflected upon meditation and solitude, nothing can be said 'that has not already been said better by the wind in the pine trees'.

RAINDROPS

'I think that I cannot preserve my health and spirits, unless I spend four hours a day at least – and it is commonly more than that – sauntering through the woods and over the hills and fields, absolutely free from all worldly engagements.'

Henry David Thoreau

I hid my bike in a hedge and set out to explore the grid square on foot, keen to see what the world would offer to my imagination today. My voice sounded small as I hummed a song to myself beneath an outsize bridge rumbling with overhead lorries. The solitude I experience in nature is very different from the loneliness that often keeps me company in daily life on this map. It is one of the appeals of swapping Wi-Fi for wellies and going for a wet schlep through muddy, empty landscapes.

My outings had recalibrated my preconceptions about how much of my map was covered with concrete and shown that much still lent itself to vigorous, solitary hikes. Though moulded by man – absolutely – the countryside is not irrevocably wrecked, even in this crowded corner of the country.

I was sheltering beneath the bridge from rain, hoping it would ease. Old habits die hard. Back when I was cycling around the world, I spent a lot of time hiding under bridges from rain or heat. Under today's bridge there were piles of nitrous oxide canisters around my feet, shiny and slippery like ball bearings. This gloomy underpass, miles from a town, had been someone's chosen spot for getting high.

I unscrewed my Thermos flask and poured a piping-hot cup of coffee down my throat. That's my drug of choice these days. Hot black coffee on cold grey mornings also brings back memories of the open road. Instinctively, I began to sing,

'One more cup of coffee for the road,

One more cup of coffee 'fore I go.'

I gave up waiting for the weather, returned the flask to my bag, pulled up my hood and got going. Away from the motorway, there were no buildings or roads on today's square, no walls or people. Just a long, muddy path crossing a dozen flat fields marked by hedges.

The hedgerows felt empty, until a chirping cloud of long-tailed tits livened things up, undulating and noisy as they flew by, gossiping and jostling, more tail than bird. The hedge was dotted with the red globes of rosehips, and the motorway sounded like a rushing river. These tiny birds build fabulous nests from moss. They are dome shaped, camouflaged with cobwebs, and then lined with up to 1,500 feathers to make a cosy home to return to on cold winter days like this.

Their nests may have been too camouflaged for me to spot, but I did find an old football in the hedge, deflated and tattered. More mysterious was that it was lying on top of an even older ball, almost buried

now beneath ivy and twigs. Losing one ball was unfortunate, but losing two was surely carelessness.

A wine bottle lay on a drain cover beside a soft heap of rotting windfall apples, like a muddy rendition of Cézanne's *Basket of Apples* still life. The web address on the drain informed me that Clark-Drain is a family-owned company, 'proud of our heritage (since 1963), which began setting new industry standards as pioneers of the first steel cover in the UK. Our core strength is our dedicated employees who embrace a clear set of values which drive everything we do. There [sic] aim is to not just deliver the best possible construction products but to help you to build a better everyday life for people, vehicle use and function of the built environment.'

I also learned from their website that this was not a mere drain I was contemplating. It was a 'ductile iron kerb gully', designed for pavement kerbs next to carriageways. It could be accessed for rodding or cleaning purposes through a wide opening area that facilitated maintenance access to the sewer system. And that is all I know about drains.

I spied several more wine bottles dumped in the bushes. There was a lot going on in this hedge. Intriguingly, they were all Pinot bottles. I tried to imagine the sort of litterer who stops on a country lane to discard the evidence of a heavy night and a preference for a very specific grape.

What else kept me company out in the rain and the mud once I left the well-stocked hedge to roam the fields? I saw a jay, a pair of magpies, plenty of pigeons (as always), and a couple of partridges that had so far evaded being shot. Dozens of sheep grazed on a single lime-coloured pasture of grass that gleamed among all the stubble fields and dark hedges. These were the sum of my companions for the morning, beyond, of course, the staggering wealth of life beneath my feet.

I walk over soil almost every day, but rarely crouch down to rub some between my fingers and give it any thought. Soil is more than the earthy home to the worms, of which the great naturalist Gilbert White noted, 'though in appearance a small and despicable link in the chain of nature, yet, if lost, would make a lamentable chasm'. It is vital to much of life on earth.

Almost all the world's food depends on soil. It cleans our water, prevents flooding and protects against drought. It captures colossal amounts of carbon – more than trees – and it regenerates in a miraculous, self-sustaining cycle.

Every cubic metre of soil contains hundreds of thousands of creatures, and a single teaspoonful has a kilometre of tangled fungal filaments. We take the soil shockingly for granted, but an ancient Sanskrit text cautioned that, 'upon this handful of soil our survival depends. Husband it and it will grow our food, our fuel and our shelter, and surround us with beauty. Abuse it and the soil will collapse and die, taking humanity with it.'

Today's empty expanse of flat, open fields emboldened me to wander at will, for the ground had not yet been ploughed or sown. In the fog and gloom, the muted greens and greys felt like being immersed in a rural version of a Rothko abstract. At one point I even took a compass bearing with the app on my phone, the first I had ever taken within earshot of a motorway and surely the only one I would need on this map! I hadn't resorted to anything so extreme back when I walked a lap of the M25, looking for adventure and wildness on the outskirts of London.

I followed my bearing due west towards a large sycamore tree until I intersected the footpath again. The wind was strong, and heavy rain lashed against my waterproofs. I chose to enjoy it because the alternatives were miserable. The Dutch word *uitwaaien* means 'out-blowing', and describes the urge I often experience to get outdoors and clear my head with a walk in bracing wind. Sloe berries shook on the hedges and trembled in the breeze. I used a hedge as a windbreak alongside a vast pile of horse manure, the third such steaming mountain I had seen today. Tractors must drive out here to dump the waste from a stable yard.

The headache of dealing with dung nowadays is minor compared to what it used to be. By the late 1800s, cities were drowning in horse manure. London alone had 11,000 horse-drawn carriages, plus thousands more horse-drawn omnibuses, each of which needed twelve horses a day, not to mention all the tradesmen's horses and carts. Each of these animals produced up to 16kg of dung per day, with a lot of mess, flies and disease following along behind. *The Times* predicted that 'in 50 years, every street in London will be buried under nine feet of manure'. In 1898, New York hosted the world's first international urban planning conference, at which they discussed the manure problem but failed to find any solutions. Was modern urban civilisation doomed?

Then along came Henry Ford with his affordable Model T, and by 1912 the impossible poo problem was nothing but a nostalgic memory. Hooray! Civilisation was saved. Today, the 'great horse manure crisis of 1894' is an analogy for supposedly impossible problems being solved by new technologies. Three cheers for cars!

The manure crisis was solved by the invention of petrol-driven vehicles, but they now play a major part in today's climate crisis. We live in a tangled web of unintended consequences, unimaginable futures, and it sometimes feels hard to do anything that doesn't cause harm.

Typical humans: we do everything to extremes. Fending off starvation through farming and ending up with more people dying from overeating than hunger. Harnessing the power of fire and eventually throwing the planet's climate into chaos. Sheltering from rain so effectively that children now spend most of their playtime indoors in front of a screen, enjoy less time outside than prisoners, and can identify more Pokémon species than real ones. What, I wonder with hope, will be our solutions to 'the great multiple crises of the 2020s'?

I distracted myself from these gloomy thoughts with another cup of (climate-damaging) coffee from my flask. I watched raindrops land on a lichen-covered branch, gather, grow and then fall.

The largest water droplets ever recorded were almost a centimetre across, observed in clouds above Brazil. That is about as big as raindrops can become. Droplets are more than just water, for cloud vapour first needs something to condense around. Each droplet forming in a cloud has a tiny particle called a 'condensation nucleus' at its centre. Without these nuclei, there would be no rain. Ash particles from forest fires explained Brazil's record-breaking raindrops.

Initially, a raindrop measures less than 0.005mm and is almost spherical. As it collides with other droplets, they merge and grow until they're heavy enough to fall from the cloud. As they descend, wind resistance and surface tension flattens the bottom of the raindrops so they become shaped more like kidney beans. Once a droplet is larger than around 5mm, the air pressure then usually overcomes the surface tension and splits the raindrop into two smaller droplets. I seemed to have spent a lot of outings on this map being doused by such marvels of nature.

The grid square's only footpath ran straight and true, a smear of white chalk through the brown mud. My map showed it heading tantalisingly towards an Anglo-Saxon burial mound, but that lay in a neighbouring square. I'd have to wait for another occasion to return. Yet knowing how way leads on to way, I doubted if I should ever come back, tempting though it often was to follow my nose onto the next square, and on and on and never look back. There were a lot of squares for me to explore before I began covering any adjacent ones.

Despite the weather and the noisy motorway, this bleak area was surprisingly charming in a stark, monochrome kind of way. An empty square, but teeming with invisible life beneath my feet, and a cold rain that reminded me to savour every sip of hot coffee. My boots were heavy with mud and my nose was numb. The wind whipped and moaned in the telephone wires and my hands had become claws. I was definitely earning the gorgeous summer mornings to come. Yet I enjoyed today once I was out there. The square offered a sense of harsh, wild solitude that I hadn't anticipated finding, but certainly welcomed.

ENCLOSURES

'There are no unsacred places; there are only sacred places and desecrated places.'

Wendell Berry

Rain was still falling hard a week later when I cycled past a garden with two life-size sculptures of giraffes, towards a modern red-brick Catholic church. On the church wall was a statue of a bored-looking Saint George stabbing down at the dragon with about as much enthusiasm as a community service litter picker. Old George up there took quite the journey to sainthood in rainy England from his beginnings as a soldier in the Roman army. He is the patron saint of not only England but also Georgia, Ethiopia, Catalonia, Aragon, Valencia, and Corinthians FC in São Paulo.

I turned right at the church into a maze of terraced streets. In one house I saw a dozen trophies shining in an upstairs window. A child's bedroom, I guessed.

An enormous railway embankment towered above the houses and overshadowed the streets. It led onto a viaduct whose ten arches were visible from grid squares for miles around. I was understanding the lie of the land better now, getting a clearer idea of how all these places fitted together.

Passing beneath the viaduct, I turned onto a footpath along the river. The stream was only small, but it was swift from weeks of rain and close to overflowing. Although the day was cold and wet, there was, I thought optimistically, a little more birdsong than last week. My bike clogged with mud as I slipped and slid along the riverbank. A cloud of winter gnats danced in the air like will-o'-the-wisps, making the most of their single week of life.

The first shoots of early daffodils pushed through the wet earth, and wrens bustled around the base of hedges. With a population of 11 million pairs, they are our most common birds (discounting 145 million chickens), which sounds surprising because they lurk and skulk in dark crevices. Their Latin family name is *Troglodytidae*. Wrens are not particularly noticeable until you learn their song, but once you do, it is unmissable. I tuned in to it once someone told me that they end each loud riff with a rat-a-tat-tat sound like machine-gun fire.

A quick wren digression (ren digwression?) The Irish word for wren comes from *dreoilín*, or trickster. They've had a reputation for cunning, so Aesop's fable goes, ever since all the birds first gathered to choose their king. After much argument, they agreed that whichever bird could fly thc highest would become king. The tiny wren stowed away in the feathers of the mighty eagle, which soared far higher than any other bird. As it descended to claim the thrown, the eagle heard a small but loud voice chirping upon its back, 'I am the king, I am the king.' The eagle refused to accept the stowaway's victory, insisting that it had

won fair and square. But the wren retorted that though the eagle might be strong, it did not have the wisdom to lead. And so the wren became the unlikely king of the birds.

A row of terraced houses, built with the viaduct and railway station 160 years ago, had long, narrow gardens reaching out like tendrils to the riverbank, wide enough only for a patch of decking and a barbecue. The ploughed fields opposite were bare and black, sprinkled with white flint. In other fields, hop poles stood tall like washing lines. In the summer they will hang with spindly vines for producing beer.

I rode across a sodden recreation field, dotted with dog shit and empty except for a solitary, wonky football goal in an island of mud – minus its net – and a lone swing missing its seat and chain and therefore its entire purpose.

Recreation fared better elsewhere on the grid square, with a bowls club (the car park sign had been donated 'in memory of Alan') and a football ground whose sheltered viewing area had been brightened up with a wall of colourful graffiti. Someone had left a bag of fizzy drinks and Jammie Dodger biscuits there (named after Roger the Dodger from the Beano comic). I was tempted to nick them, but instead had a play on the outdoor gym equipment installed by the pitch.

Returning to my bike, I found a feisty robin perching on my handlebars and singing lustily. Its red breast burnished the grey day, though to my eyes it looked more orange than red. The word 'orange' to describe a colour didn't exist in English back in the 15th century when it became popular to give creatures human names, hence 'robin redbreast'. Only when the first sweet orange trees arrived in Europe in the 16th century, with their startling fruit, did the term 'yellow-red' seem inadequate for that brilliant shade, and a new colour was born.

I find it interesting that describing colours is not a simple matter of black and white. Chaucer described a daisy as being white and red. Russians and Greeks separate what we would call light blue or dark blue into two distinct colours. Irish and Turkish languages differentiate between different reds. Green is often considered a shade of blue in Japan, and the Pirahã language in Brazil has just two colours: light and dark. I waited for the little red/orange-breasted bird to leave my bike, then schlepped onwards through the mud.

The bubbling river had been the catalyst for a community settling here long ago. The stream powered small flour mills, generating enough customers to also keep a forge in business. A paper mill was built in the 19th century, whose brick chimney no longer billowed smoke but still towered over the town. It must have looked so impressive when it went up. There was an astonishing energy in the country back then. Yet it also reminded me how quickly eras come and go in the grand scheme of things. Today's cutting-edge innovation is tomorrow's anachronism and next week's museum piece. The mill had now been converted into apartments for commuting millennials, with a strong whiff of weed drifting from a second-floor flat.

This square promised a lake as well as the river. But high hedges, 'Private property' signs and a spiked fence foiled my attempts to get to the shore. I shook the railings in frustration. Most of the problems in my life, small though they are, are connected to not spending enough time in wild places. And I am one of the lucky ones: I am writing a book about nature and I earn my living from playing around outside. Not everyone gets to be so engaged with the world around us.

There is a glaring problem when 90 percent of adults want their children to learn about wildlife, but half cannot recognise a sparrow, and only 1 percent of families can identify our most common trees. Most of us feel nature is important, yet we don't spend enough time in it, and being distanced from nature affects people's health and happiness.

The lake had been made by flooding an old gravel quarry. It was secured as fiercely as a prison but looked lovely through the chain-link fencing. A few fishermen sat around the shore, private members hunched beneath green umbrellas and surrounded by paraphernalia to outwit the lake's carp, perch, chub, tench, roach, bream, eels and dace.

I met a fisherman by the exit as he was leaving with his wheelbarrow of equipment, locking the tall gate behind him. I asked him if the occasional loud bangs I heard were intended to scare herons away.

'Cormorants,' he answered. 'Horrible things. They've moved in from the sea and taken over. We had eighty of them on the lake once. They'll go for anything, even fish too big for them to eat. They need to be culled.'

All of which made my head spin a little. What is the best way to 'control' wildlife? Do we heed the bird lovers or the fish lovers? To cull or not to cull? To evolve or to preserve? Does prior presence give priority? The only thing I was sure of was that when humans get involved, nature usually gets messed around.

I moved on from cormorants to ask about the fences. 'I wish I could walk round the lake, but why is it all fenced off?'

'I don't like it either,' he sympathised. 'But we need to do it, sadly. Sometimes I love to just go for an evening walk round the lake. I don't even bring a rod. Just enjoy it. It all used to be open, but it was chaos. Dog walkers and anglers don't really get on. Crap everywhere. Dogs running around, noses in the bait boxes. Jumping in the water. Getting hooked in the mouth. Chaos! But it still worked, just about. It was all open until, oh, less than twenty years ago.'

'Then what happened?' I prompted.

'I'm not racist, but…'

My ears pricked up. What did fishing have to do with race?

'But it was poaching that did it. When all the Polish people came here about twenty years ago. Nice people, don't get me wrong. But they eat carp. They even eat it for Christmas dinner.*

**Carp has long been the traditional Christmas Eve feast in much of Central Europe. Families keep a live carp in their bath for a couple of days before eating it, to allow all the mud to be flushed from the bottom-feeding fish's guts. Plus you have the bonus of a new family pet, until you eat it. They soak the carp in milk before cooking to improve the flavour and make it less 'fishy', then carve it vertically to create horseshoe-shaped fillets to bring good luck. The fish scales also represent good fortune and some people keep one in their wallet for the year.*

'You see them all down in the town,' he continued. 'Catching fish under the road bridge, slipping them into their bags. Our lake was getting rinsed by poaching. So we had to put up all these fences. If you want the real problem, though, you've got to blame the Enclosures Acts, don't you? That's when all this began. Hundreds of years ago.'

For centuries, English agriculture operated a system of common land. Some land was privately owned, but people still had access to it. The word commoner originates from someone who benefitted from these commons.

The idea of ownership equating to exclusion only came later. Although there were specific rights, everyone could use common land for grazing, foraging, gleaning (gathering leftover grain after the harvest), and sometimes fishing or hunting. Farming was based on large, open fields where yeomen and tenant farmers cultivated strips of ground alongside each other. It was not an easy life, and rural poverty was severe, but it was relatively fair and sustainable.

Then, between 1604 and 1914, Parliament passed more than 5,000 acts covering open fields and common land affecting 20 percent of the area of England. The new laws allowed landowners to enclose their land and prevent commoners using it. The acts were motivated by landowners looking to maximise rental income. They altered our country's relationship between the people and the land for ever. Farm workers suffered from these rents, and many were forced to leave the ancient commons to seek work in towns and cities.

The other side of the argument was that consolidating holdings led to more economical farming systems. Landlords could introduce innovations, which led to the agricultural revolution, increased productivity and boosted profits. Property rights had been a Roman idea, resisted by both Greeks and Celts, who couldn't see how nature could be owned by humans. Centuries after the Romans left Britain, however, the Enclosure Acts tilted the balance of power, ownership and access further away from the common man than ever before.

Today, half of England is owned by less than 1 percent of the population.

So I now found myself peering over a fence at a little lake I'd have liked to paddle in, but which was somehow 'owned' and therefore off-limits. I murmured an 18th-century folk poem to myself, which protested against the enclosures that started all this:

'They hang the man and flog the woman
Who steals the goose from off the common,
Yet let the greater villain loose
That steals the common from the goose.'

IMBOLC

FEBRUARY

COPPICE

'It is remarkable how little we attend to what is passing before us constantly, unless our genius directs our attention that way.'

Henry David Thoreau

'Right, Humphreys. Stop procrastinating. You haven't even started yet!' I rebuked myself, and stepped out into the rain to begin. As always, *solvitur ambulando*, I solve things by walking.

It is the gloom that does for me in winter. Seven of my past eight grid squares had been grey and wet, in a winter where the rain never ended and the sun never shone. I was sagging like a feeble houseplant, pale and etiolated from lack of light. If I could hibernate until spring returned, I would.

I dearly wished to dig out my passport and head somewhere far away where sunlight shone hot on my back. California called. Emigration enticed. Marrakesh, maybe? I find the dark half of the year harder to endure every year. But just when I am about to crack, I recognise tiny changes heralding the approach of spring and the return of all good things.

So it encouraged me to hear a definite increase in birdsong this morning, a ratcheting up of woodland activity. Perhaps life was returning, and perhaps my own life was too. For today was Imbolc, the Gaelic festival celebrating the onset of spring that occurs halfway between the solstice and the equinox.

The word Imbolc comes from the Old Irish *i mbolc*, 'in the belly', referring to lambing season. It has been celebrated in Ireland since Neolithic times, with some tombs aligned to Imbolc's sunrise and the usual excellent celebrations revolving around fires and feasts.

I walked through a motorway underpass that led into the grid square like a portal into today's explorations. The echoing tunnel had been decorated with spray-painted murals, including a blue tit in the branches of a sweet chestnut tree, a beetle under a magnifying glass, and a soaring albatross.

An albatross aloft is a spectacular sight, for its eleven-foot wingspan is the greatest of all birds'. It allows wandering albatrosses to soar and glide for thousands of miles over the Southern Ocean. Confined to my small map, in this rainy, restricted season, I envied their freedom. As well as the usual scrawl of graffiti tags and swearing, there was an all-caps exhortation to 'DO A BETTER JOB'. I resolved to get my curiosity hat on and to do just that.

A landmark of each new year for me is when the snowdrops appear, once called Candlemas bells, Eve's tears, February fair-maids or Mary's tapers. Bunches of the little white flowers were popping up among the

green leaves of dog's mercury that covered the floor of the country park woodland. The 'drop' in the name refers not to snow but to the flowers' shape, like an 'eardrop' or earring. I'm not sure if my fondness for them qualifies me as an out-and-out galanthophile, a lover of snowdrops, but I do like them. They are relatively recent immigrants to our shores from southern Europe, arriving only four centuries ago and taking another two to establish themselves as wild plants. How long must an immigrant be here before they belong and we claim them as our own?

In Romania, it is said that the sun was a beautiful maiden who returned to warm the world after winter. One year, winter kidnapped the sun. A hero fought winter to free the sun, saving us from eternal frost. Winter mortally wounded the hero, but as drops of his blood fell and melted the snow, snowdrops grew from those droplets.

In Germany, they tell the tale that God asked all the flowers to donate some of their colour to the snow. All refused except for the generous snowdrop. In return, it got to be the first flower of the year to bloom.

Snowdrops traditionally emerge on Imbolc, and legend suggests they've been flowering then ever since the Garden of Eden. Tiring of the endless winter after her banishment from the garden, Eve was visited by an angel who created snowdrops out of snowflakes to prove that winter does not last for ever. They have symbolised hope ever since, and I share that feeling as the earth wakes up towards spring once again.

I sat on a wet bench in my waterproofs and looked around. Most of the tree branches were bare, except for the crispy, rustling leaves of a few small beech trees that had held on to their withered leaves through winter, known as marcescence. A woodpecker rattled nearby as the rain rattled on my hood, and I listened to a noisy great tit calling over and over, 'teacher, teacher!' Rosehips offered some colour, each red berry hung with a shining raindrop like Vermeer's pearl earring painting. Colourful, too, were the yellow winter aconites and the flowers of a prickly gorse bush. Gorse used to be gathered from common land for fuel, fodder, making floor brushes and chimney brushes, and also for dye to paint eggs yellow for Easter festivities.

While I sat on the bench, a steady trickle of dog-walkers passed me, swinging long-armed ball launchers. The dogs boasted a quirky variety of names and were of all shapes and sizes, like their owners. I wondered when we became so obsessed with pets. Humans have lived alongside animals for millennia, but beyond an occasional elite extravagance, people only began keeping pets in the 18th century. An evangelical zeal towards raising moral offspring led to a number of books encouraging children to rear small animals to help them learn kindness, commitment and practical nurturing skills.

Working-class families in cities valued birds for the colour and song they brought into their cramped homes, an early example of yearning for nature in an increasingly urbanised and atomised society. Keeping songbirds became so popular that legislation was introduced in the 1870s to limit the number captured, although a wild bird market persisted well into the 20th century.

Today there are 24 million pet cats and dogs in Britain, with almost a quarter of households having a dog. The world's pets now weigh about the same as all the planet's wild animals. Pets are lovely, but they are yet another accidental nail in the nature coffin.

Farmers striving for higher-level stewardship by creating fallow field margins for wildlife are often disheartened by dog poo and dog walkers trampling the ground. Dogs allowed off leads reduce bird numbers and biodiversity in an area by more than 35 percent. And our cats kill more than 200 million mammals, birds, reptiles and amphibians each year. Sir David Attenborough has argued that all cats should have to wear bell collars, while Australia is considering cat curfews to protect native wildlife.

Our beloved pets also eat vast amounts of meat, for which worldwide production requires a land area twice the size of Britain. America's cats and dogs alone generate the same greenhouse gas emissions as 13.6 million cars.

February may be a drab month, but a few flowers do still bloom, including hazel trees with their thousands of yellow catkins. The name derives from a Dutch word for kitten, since the catkins resemble fluffy tails. In some places they are called lambs' tails, a more obvious connection to spring, perhaps. Catkins are unusual-looking, dangling flower clusters (there are actually 240 flowers per catkin), which produce terrific amounts of pollen. They rely on the breeze for dispersal and then hope that at least one grain from the cloud of billions of microscopic pollen grains will land upon a female hazel flower.

Until now I had selected each week's grid square at random. But today I had cheated and chosen where to go with deliberation, looking for an environment to cheer me up. I had picked woodland, my current favourite landscape. Yet a blob of green on a map doesn't do justice to the land it represents. That's why we need to go out and explore for ourselves, I suppose, rather than resting contentedly in the world of maps. Despite promising, ancient names on my map – 'Broad Oak Wood', 'Great Wood', 'Birch Wood' – the woodland I walked through today had been viciously coppiced and looked like a war zone.

Coppicing involves felling trees on a rotational basis and then allowing new trunks to regrow from the base. These grow more quickly than replanting, as the trees already have established root systems and are less vulnerable to grazing, shading or drought. Trees can be coppiced indefinitely, and it is a technique that has been used for thousands of years to provide timber and firewood. Coppiced wood was used in 3807 BC to construct a Neolithic track across boggy ground in Somerset, known as the Sweet Track.

The technique works with most species, but the most commonly coppiced trees are hazel, sweet chestnut, ash and lime. Woodland is often managed as coppiced-with-standards, where scattered trees are left to grow uninterrupted into mighty 'standards' before eventually being felled for timber (traditionally for building houses or ships), while the understorey is coppiced every couple of decades.

Although it appears destructive, rotational coppicing promotes biodiversity and slows the spread of invasive species such as rhododendrons and grey squirrels. Coppicing small areas at a time generates patchworks of open spaces and scrub thickets, as well as broadleaf woodland. Scrub is so important for the growth of new woodland that in the 18th century you risked three months' hard labour and lashes of the whip if you damaged it. There is an old saying that 'the thorn is the mother of the oak'. Long ago, large herbivores such as red deer, boar and auroch (extinct wild oxen) kept woodland clearings open. But since we killed them off, coppicing has become a severe but necessary part of woodland management.

Without coppicing, woodlands become uniform, with few clearings. The full canopy starves smaller plants of resources, resulting in little biodiversity beneath the trees. The successful trees are all of a similar age, so when they die at roughly the same time, it leaves a barren landscape that has to begin again. Coppicing keeps a wood healthy, thriving and varied. Many species would struggle to survive without it.

When an area of woodland is coppiced and the trees are felled, new plants germinate and flower in the sudden abundance of sunlight. Their seeds may have lain dormant since the canopy last closed overhead, or drifted in recently on the breeze. Insects and butterflies are attracted to these open areas, and birds and bats follow along. A few years later, as the coppiced shoots regrow, the area is tangled with fast-growing brambles, bracken and honeysuckle, generating havens for muntjac deer, foxes, stoats, weasels and nesting birds.

Once the coppiced trees reach about twenty feet, they start to outshade and out-compete everything else again. Songbirds nest in the branches and the ground below thins out. After twenty years, the trees are ready for felling once more, and the cycle continues.

The woods today were not currently as scenic as I had hoped when browsing my map. A sizeable chunk of today's walk felt like I was walking through an Amazonian clearcut. Even so, I was aware as I continued towards the end of the square that I had hiked myself happy. Shoots of recovery had replaced my murky reluctance at the start of the day. There was new life in these coppiced woods, and we would both soon rise and flourish again.

SNOW

'We too have our thaws. They come to our January moods, when our ice cracks, and our sluices break loose. Thought that was frozen up under stern experience gushes forth in feeling and expression.'

Henry David Thoreau

There are two types of people in the world: those who love snow and those who do not. There is no such thing as a child who does not like snow. A few people have valid objections to snow: those with broken hips, and confounded commuters, for example. But anyone else whose heart does not leap at the first falling snowflakes is a miserable curmudgeon. There, I've said it! I get as excited by snow today as I did back on those glorious, rare occasions at school when someone in the classroom yelled, 'It's snowing!' and cheery pandemonium broke out.

The south of England being a mild sort of place, the best I hope for each year is a covering of a few inches, a couple of sledging outings and a day or two of jolly disruption. Today, after weeks of rain, I was excited to get out into the snow that had fallen overnight, not least of all because I had also noticed the dawn arriving a little earlier.

It was nice to get away from the daily grind of book-writing in my shed. Snow makes everything feel more adventurous, though the

sprinkling here couldn't compare to the majesty of hauling a sledge across Greenland's vast silence, relishing being self-contained with a couple of friends and very far from civilisation. But I was still thrilled.

'As the days lengthen, the cold strengthens' goes the proverb, with a nod to scientific veracity. Earth receives its least sunlight at the winter solstice, yet the coldest temperatures come later, a seasonal lag caused by more solar energy leaving the atmosphere than arriving. Today, the snow muffled the world and quietened everything. I could hear a buzzard and the cawing of rooks, but the usual motorways and sirens sounded softer.

I paused in a field to admire a newly laid hedge filled with a healthy variety of species, including hawthorn, blackthorn, hazel and maple. We destroy thousands of miles of hedgerows every year, so this conservation work was gratifying to see. Hedges play an important role, not only for aesthetics and tradition, but also for shelter and food, and as movement corridors for wildlife, including insects that pollinate crops and birds that eat pests on those crops.

The hedgerow's trunks and large branches had been sliced partway through, then bent over and woven through each other. It looks severe, but over time, the hedge grows back into an impregnable barrier. Laying hedges is a country craft that has been practised for hundreds of years, as I learnt from the niche National Hedgelaying Society.* Hack too much off a hedge and it degrades, becoming hollow at the bottom

**Hooper's Rule can tell you approximately how old a hedge is. Its age in years equals the number of woody plant species in a thirty-yard section, multiplied by 110.*

and useless for sheltering wildlife. Yet if hedges are ignored, they grow too tall and collapse. 'Hedgelaying preserves the past and protects the future,' the society therefore declares, proudly.

As I followed the hedge, I saw the tracks of a pickup truck in the snowy fields, the swish of a sledge, the gait of a runner, the hop of a rabbit, the strut of a pheasant, and the heroic efforts of a mole hauling its way through the cold soil beneath us and excavating an impressive row of fresh molehills, black against the white field.

Moles are the only mammal to live solely underground. They survive in that low-oxygen environment by having high numbers of red bloods cells, and they certainly cope with the darkness better than I do in these long winter months. They are light sensitive and not blind, but they do also have sensory hairs over their body to help navigate in the dark. Their ears are inside their bodies, behind their shoulders, and their snout acts like a sound tube. They are tough little beasts. Despite being less than 15cm long and weighing under 150 grams, they can still dig 200 metres of tunnel a day, shifting 540 times their own body weight of soil. That's like me shunting eleven hippos around. Moles work for four hours, then rest for four hours, in a tough underground shift system that continues all day, every day, eclipsing my mere forty-five days and nights of shift work when rowing the Atlantic.

All this exertion means that moles need to eat half their body weight a day. For me, that would equal 350 quarter-pounders (or half a kangaroo); for a mole it means around twenty worms. If they struggle to find enough in their existing tunnel network, they dig new runs, which means new molehills and potential conflict with farmers.

Some farmers try to get rid of them because they believe moles pollute silage, wreck pastures and encourage weed growth. Mole catchers still exist, but back in their heyday every parish employed one. They earned more than surgeons, plus they also sold the silky pelts to be tailored into waistcoats. Secret mole-controlling techniques were carefully handed down through families over the generations.

The bitterly cold winter known as the 'Little Ice Age' in 1566 led to mole control becoming national policy in an attempt to protect food supplies. Queen Elizabeth passed 'An Acte for the Preservation of Grayne', which remained law for 300 years. It included bounties for the destruction of 'vermin' that included everything from hedgehogs to kingfishers and, most certainly, moles.

I was walking through the snow towards the northern edge of the day's grid square. Bare branches stood out on a lone tree like the veins and vessels of a heart, silhouetted against the low sunlight and framed by a dark and heavy snow front moving my way. I enjoy watching weather approach like this. The sky was hazy, as though draped in gauze, and all that falling snow was soon going to reach me. I hoped for those jumbo soft snowflakes that fill the sky and cover the land with magical six-pointed stars.

The uniqueness of each snowflake is one of the best-known scientific facts, and the astronomer and mathematician Johannes Kepler was perhaps the first to pay inquisitive attention to this natural phenomenon. In 1611, he made a booklet as a Christmas gift for a friend, titled *The Six-Cornered Snowflake*, pondering explanations for their shapes. It may have been an idle diversion from his revolutionary study of the planets, which changed our understanding of the cosmos, and from the trauma of his mother being accused of witchcraft and spending more than a year chained in a prison cell.

As the weather front reached me and it began to snow, the fields were soon covered in what looked like beanbag pellets. From this mystery, I learnt a new word. These pellets were 'graupel', formed in a process called 'riming', when supercooled drops of water (water that remains liquid at temperatures below zero) freeze around existing snow crystals in the sky.

I walked on, wondering whether I might spot all eight official categories of snow: column crystals, plane crystals, a combination of column and plane crystals, aggregation of snow crystals, rimed snow crystals, germs of ice crystals, irregular snow particles, and other solid precipitation.

Trudging through the white world felt like taking a trip back to slower, quieter times. I saw a church tower in the distance as rooks swirled, black against the white sky. Partridges flew by in a whirr of short, fast wingbeats, no pear trees in sight. A woodcock burst from cover. There was less to see in this blanketed grid square, but there were abundant things to feel.

I gave thanks for my modern winter clothing (especially the home-made scarf that I was very proud to have knitted myself), and savoured my flask of soup, particularly as today I had some of my home-baked bread to accompany it.*

Natufian hunter-gatherers began making bread from wild wheat, barley and plant roots 14,000 years ago in Jordan. Bread spread quickly across the world once wheat was cultivated in the Fertile Crescent, and it played a significant role in the establishment of settled towns in preference to nomadic lifestyles.

The *Deipnosophistae*, a 3rd-century book about a series of Roman banquets, contains recipes for griddle cakes, honey-and-oil bread, mushroom-shaped loaves covered in poppy seeds, and the military specialty of rolls baked on a spit that continues down the ages to the charred but enjoyable campfire bannocks made by Cub Scouts today.

My bread was so easy to make, cost only pennies, yet tasted so blooming delicious that I scoffed almost half of it, sitting on a log among a billion falling flakes of snow, and catching snowflakes on my tongue like a kid who has never fallen out of love with snow.

**Here are two easy recipes I use:*
www.bakerbettie.com/easy-no-knead-skillet-bread
www.leitesculinaria.com/99521/recipes-jim-laheys-no-knead-bread.html

MARINA

'And this our life, exempt from public haunt, finds tongues in trees, books in the running brooks, sermons in stones, and good in everything. I would not change it.'

William Shakespeare

Historical names for February include the unappealing Old English terms *Solmonath* (mud month) and *Kale-monath* (cabbage month). I'm pretty sure the Hawaiian word for February, *Pepeluali*, refers to neither mud nor cabbage.

In Finland, February is *Helmikuu*, the month of the pearl, when snow melts on branches and forms droplets. In Poland it is *Luty*, meaning freezing cold. For Macedonians, it is the month of felling trees and chopping wood, *Sechko*.

I was definitely in a mud month here, rather than a pearl month. The mountains may have been calling, but they felt farther away than ever. I had one of my occasional bouts of wondering if I should give up limiting myself to this single map.

'What a stupid idea all this is,' I moaned. 'It's cold. It's wet. I could either be in bed or in Bali.'

Last week's snow had melted, and the world was now drenched and dull again. The month of felling trees felt relevant as I arrived in today's grid square to find swathes of cleared land, stripped and flattened ready to build more flats and houses.

Britain has a chronic housing shortage and building rates have almost halved in fifty years. The government therefore pledged in its most recent election manifesto to build 300,000 new homes a year (before giving up on that promise). We need more houses. But on the other hand, land is a finite resource, and my map feels destined to disappear beneath industrial parks and cul-de-sacs. We need homes with green areas and space to stretch our legs, but low-density, sprawling developments also assume that there is infinite space to spread out across.

An orange digger was carving away the earth around one of the few remaining tall trees, leaving it marooned and isolated. Scoop, spin, dump, repeat. Scores of saplings had been planted along a verge, but they were so crowded in their plastic protective tubes that few would reach maturity. It was a quota-filling exercise to say 'we planted hundreds of replacement trees', rather than any meaningful compensation for the uprooted land.

However, the new policy of biodiversity net gain (BNG) does compel developers to try to avoid habitat loss. If this is not possible, they must recreate the habitat, plus 10 percent extra, elsewhere or buy statutory credits from the government as a last resort. Such systems are still flawed and evolving, but this is an exciting time for conservation. We have caused so much destruction, but we finally live in a time with an ever-growing appetite and momentum to leave the environment in a better state than we found it.

The builders had fenced off the footpath along the river for the next four years and sent me on a detour down a road instead. I had another flush of irritation that this project was pointless. It frustrated me to be stuck in the crowded southeast of England when I pined for wilder landscapes. But if I don't explore close to home, and if I neglect local landscapes in favour of exotic places, it is a further slide into being unaware about nearby nature, and then not caring or being aware about its demise, and so doing nothing to call it out.

Needing a fresh perspective, I changed direction and climbed down onto the stony river beach in front of a block of new apartments. I laid my bike down and crunched over the pebbles to the water's edge. The river was wide and swift. The sky and the water were the same sulphurous hue, like cold tea.

Pootling up and down the shore calmed me. I carefully observed the gentle lapping of the water and the colours of the beach. Dark flint nodules, white lumps of chalk, cracked red bricks, green slime, glossy black seaweed and a child's pink Raleigh bicycle in the grey mud. A flock of redshank, dappled brown with long red legs, probed the shallows with their long bills. I indulged in a spot of mudlarking, finding a steering wheel and a football among the usual flotsam and jetsam of broken bottles, flip-flops, traffic cones and shopping trolleys.

A pulsing mass of sandhoppers thrived in all this junk. They look like a cross between a shrimp, a woodlouse and a flea, with semi-translucent bodies and outsize back legs. Sandhoppers are less than a centimetre long but can leap thirty times that distance with a flick of their rear end. They scavenge, shred and eat almost anything, which helps to break down decaying matter washed up on shorelines. But these amphipods also contribute to the microplastics crisis in our oceans by shredding plastic bags into over a million microscopic fragments, which then disperse through the food web and become impossible to remove.

I headed onwards, along the river's bulky flood barriers. Someone had once walked down here before the concrete set, leaving footprints as a ghostly reminder of their passing. So many footsteps have walked this way over time, the community linked to the river for thousands of years, piling new buildings onto old ones, history on top of history, always changing and evolving and beginning again. Ghosts and footprints, beginnings and endings, and more beginnings.

Take, for example, the grand house I was surprised to discover among the modern developments. Perched on a grassy hill at the end of a boulevard, it has a history reaching back to the 14th century, when the manor was home to nuns, before becoming a country house for Henry VIII. Today's owner is an oil billionaire with an African model wife and eight kids whose family featured on a reality TV show about Britain's flashiest families.

There was once a nautical training college down this way too, preparing cadets to serve in the Merchant Navy and on the famous *Cutty Sark*. In 1883, *Cutty Sark* set sail on a record-breaking voyage. In order to make the most of the trade winds, Captain Woodget travelled farther south than other commanders dared to do, way down into the latitudes of the Roaring Forties.

'Below forty degrees south, there is no law; below fifty degrees south, there is no God,' warned sailors of the age.

The ship faced gales, massive waves and frequent icebergs, but *Cutty Sark* dominated the wool trade for a decade. After loading 4,289 bales of wool and twelve casks of tallow in Australia, she turned for home, arriving 25 days faster than any other ship of the age. Australia no longer seemed so far away, and our perception of the world shifted a little. Having both sailed and rowed across the Atlantic, I can imagine how this sudden shrinking of great oceans felt like a seismic change in the order of things.

On this built-up square, I hadn't thought to look for the symbol for cliffs on my map, but there they were. The jagged lines marked some white chalk cliffs in a wooded park by the river, dotted with follies that had become homes for a population of pipistrelle bats. Although a common pipistrelle is so small that it can fit into a matchbox, it still eats 3,000 insects in a single night. Pigeons nested on the cliff faces that were their original habitat before they discovered the vertical cliffs of urban buildings. A woodpecker drummed in a tree and thrushes sang their enthusiasm about the approach of warmer weather. I had thought this square was going to be uninspiring, but there was so much here I never knew about.

Feeling cheerful now, I pedalled through a new housing development out into a scruffy landscape of tired industrial units. Plastic flapped in the wind and snagged on barbed-wire fences, known in Ireland as witches' knickers. There was litter everywhere, crushed beer cans mostly. It was the sort of no-man's-land where you'd be smart not to hang around taking photos of the scaffolding businesses run out of Portakabins with gleaming black Range Rovers lined up outside.

The sight of a pylon beyond the small industrial estate lured me to the very edge of my grid square, and perhaps a tad beyond. But this was no ordinary pylon: it was enormous, soaring skywards over the salt marsh and dominating the sky. Its stark symmetry was appealing in its own way. I really enjoy places like this.

'Topophilia' describes a special love for peculiar spots, a form of place attachment. W.H. Auden, who coined the term, stressed that it had 'little in common with nature love' and emphasised an infusion of history and story into the landscape. I had to ride through a strange, empty edgeland to reach the pylon, the relics of an old concrete works now covered in potholed tracks with 'Keep Out' notices, scrubby thickets, grubby grass, silver birch saplings, stonechats, graffiti tags, colourful flowers such as celandines with what D.H. Lawrence described as their 'scalloped splashes of gold', and everywhere the glimmer and sparkle of broken glass like winter frost.

The 190-metre-high pylon held a 400kV cable that stretched far over the river. Its base was fortified with CCTV cameras, railings, and electric fences. It seemed an ideal model for Pylon of the Month for the Pylon Appreciation Society (life membership £15). I had intended to turn around when I reached it, but beyond the undergrowth, I spotted the tall mast of a wooden boat. Intrigued, I pressed on farther towards what was to become one of my favourite hidey-holes on the entire map.

I had stumbled upon a creek tucked away from the main river and folded into the marshy shore. It was low tide, so the creek was just a muddy riverbed lined with reeds. It was home to a fantastic ramshackle flotilla of boats. Moored among the rotting houseboats were two graceful wooden barges, both around a hundred years old. They used to carry cargo to Spain or ferry cement to Cornwall, returning loaded with granite or china clay from the 'Cornish Alps'.

'Marina' is too grand a term for the dilapidated walkways hammered from pallets, for the driftwood cabins and the curious craft in various states of what I suspected were eternal 'repairs'. But this discovery was more alluring than a shiny, impersonal marina anyway. The peeling hulls, scuttled boats and wobbly homemade jetties reminded me of the swamps of North Carolina or the bayous of Louisiana.

The only sound came from a small man with a wizardy face. He had a goatee beard and a flat cap, and was whacking a rusted trailer wheel with great determination. His young grandson patiently watched the hammer-wielding. The man told me he was renovating a little motorboat so the two of them could go out fishing. I admired the idea, but hoped the lad could swim.

'We go right out, far out. There's cod out there. Whiting. Flatties. Flounders and that.'

Most people like it when you express an interest in their place. Sure enough, between hammer blows, he shared his stories. He spoke fondly about the owls that lived on this scrappy marsh, the hovering kestrels, and the nightingales that churred their famous song on late summer nights. He lamented that all this liminal nature was the scene of a planning battle between conservationists and developers with ideas to cash in and modernise the area. It would be a tragedy if this boatyard and creek were paved over to make a car park. I said goodbye to the man and left him to his repairs, then cycled down the main riverbank to set up my camping stove, make coffee, and take it all in.

Yes, there were shopping trolleys in the mud and shocking amounts of plastic rubbish. But I was in high spirits as I sheltered behind an enormous concrete block spray-painted with 'AHOY' and a smiley face. At the start of today, I'd been questioning what I was doing. But the grid square had answered with so much history and surprising beauty. Maybe I wasn't done with this project after all.

GOLF

'The man who is often thinking that it is better to be somewhere else than where he is excommunicates himself. If a man is rich and strong anywhere, it must be on his native soil. Here I have been these forty years learning the language of these fields that I may the better express myself.'

Henry David Thoreau

This was a landmark day of the year: my first bike ride without wearing gloves. I woke to a softer, earlier, warmer sunrise and cycled out to a grid square that began on an overgrown heath of bracken, gorse and heather. It needed an auroch or two to control it. I continued into a wood whose highest branches swayed in the stiff breeze. I lay on my back and watched the twigs rattling against each other like squabbling fingers, snapping and cracking. There was a slight gap between each tree crown, like a mosaic. This 'crown shyness' is caused by the reciprocal pruning I was watching and it helps trees to remain healthy and share resources. The spacing improves each tree's access to light and can deter the spread of diseases, parasitic vines and leaf-eating insects.

A jay screeched as I looked at some oak galls growing on a young tree. You often spot jays near oak trees and their relationship is an example of symbiosis where both species benefit from the relationship.

A jay can gather 5,000 acorns in its life, and though they are its primary food source, the bird also helps the oak to pioneer new ground by burying the acorns in open spaces. This habit helped oaks spread across Britain after the last ice age, 10,000 years ago.

Acorns can germinate in woodland, but they're often out-competed by faster-growing species. By burying acorns over wide areas, jays help oaks to grow in less-competitive scrub. By the time the bird returns to eat its acorn, a seedling has often sprouted. The jay pulls it up, scoffs the acorn, and then, for some reason, jabs the seedling back into the ground. The tiny tree tolerates this brief uprooting in exchange for growing in sunny conditions away from the competitive woodland.

The oak galls that had caught my eye were caused by parasitic wasps. Galls are smooth, dark balls, like large marbles, that protrude from a tree's twigs, with a tiny wasp larva growing inside. These ones were caused by a wasp deliberately imported from the Mediterranean 200 years ago because the high tannin content of the galls was important for the leather-tanning industry. Oak galls were also used to make ink, with recipes dating as far back as Pliny the Elder in the 1st century AD, and still used as recently as the 19th century when the United States Postal Service still had an official recipe for the ink that was to be used in all its post office branches.

This was an area of expensive homes tucked into individual padded envelopes of woodland, hidden behind tall gates with security cameras. There were very few visible houses on today's square, apart from a terrace of old cottages along a lane up a narrow hill. A footpath ran down the back, and in one of the gardens I glimpsed dozens of statues of Buddhas, Chinese dragons and carved lions behind a hedge of Asian bamboo.

Bamboo flowers infrequently, though it does so in synchrony and gregariously, meaning that all the plants in an area burst into flower at once. The flowering interval of *Phyllostachys bambusoides* is an astonishing 120 years. We can say this with certainty because Chinese scholars have kept records of the phenomenon for more than a thousand years.

I followed a path past an abandoned digger whose flaking blue paint and red rust patches had acquired a natural covering of orange lichen. It created an attractive palette. As I walked deeper into the wood, the air rang out with chainsaws and diggers, and there was a strong smell of smoke. Trees were being felled and burnt everywhere to clear space for some new houses.

I crossed a busy road to get away from the destruction, whereupon the cat's eyes caught my eye, as they're supposed to do, I guess. Their inventor, Percy Shaw, was a Yorkshireman who enjoyed a pint or two in his Queensbury local, the Old Dolphin, after work. While negotiating a twisty lane home one foggy night, he was warned by the reflection from a cat's eyes that he was in danger of driving off the edge of the road.

Inspired by this, Shaw began tinkering in his spare time. He patented his invention and established a company to sell his 'cat's eyes'. His big break came with the blackout in the Second World War that made driving more treacherous and led to a boost in sales. Today, cat's eyes are used throughout much of the world. To complete Shaw's fame, he even has a Wetherspoon's pub named after him.*

Early spring is an exuberant time of year, filled with sounds of the joy of freedom. Starlings chattered merrily in an ash tree. They have a radiant oil-like iridescence, and speckles on their winter plumage, the stars that give the birds their name. We have the charming

**When I was about eighteen, I gouged a cat's eye from the road in my village and set about trying to turn it into a necklace. Fashion has never been my forte. Nor has DIY: I failed even to take the thing apart.*

'murmuration' collective noun for starlings, but they are also sometimes referred to as a chattering, a scourge, a vulgarity or a filth because of their noisy and messy habits.

Starlings are our most talkative birds, with a repertoire of up to thirty-five songs, plus bonus additional clicking tracks. They are excellent mimics, learning new sounds and then passing them on to their offspring. Henry Mayhew, the founder of *Punch* magazine, described them as 'the poor man's parrot' on account of their mimicry and plumage.

Huge murmurations of starlings, such as those at winter dusks on the Somerset Levels, are one of the greatest natural spectacles I've seen in Britain, rivalled only by the seabird colonies on the Farne Islands. Samuel Taylor Coleridge once described a scale of murmuration in London that no longer exists. 'Starlings in vast flights drove along like smoke, mist, or anything misty without volition… some moments glimmering and shivering, dim and shadowy, now thickening, deepening, blackening.'

'Starlings in Winter', by the poet Mary Oliver, also captures their starry beauty in flight.

'This wheel of many parts, that can rise and spin
over and over again,
full of gorgeous life.'

The lanes were filled with daffodils, a sure sign that spring was beginning. Mentioning Coleridge and daffodils so closely together inevitably leads to Wordsworth. In 1795, the men became friends and published *Lyrical Ballads* together, a volume of poetry said to have ushered in the Romantic Age of literature. It sold well over the years and Wordsworth earned money from poetry for the rest of his life.

Nobody took much notice of 'I Wandered Lonely as A Cloud' at first, but it became Wordsworth's most famous poem, with its rhythmic language that captures nature's peace.

'I wandered lonely as a cloud
That floats on high o'er vales and hills,
When all at once I saw a crowd,
A host, of golden daffodils;
Beside the lake, beneath the trees,
Fluttering and dancing in the breeze.'

Personally, I've always disliked the poem, ever since I had to memorise it at junior school and, on another occasion, to copy it out laboriously for a handwriting competition that I was never likely to win. But I do like the cheerful optimism of the flowers themselves. They bring colour and hope back to the land before winter has truly passed and before the warm days settle in. 'Daffodils, that come before the swallow dares', as Shakespeare wrote in *A Winter's Tale*.

Daffodils grow across Europe and North Africa. They were first planted in gardens around 300 BC, and the enthusiastic botanist and philosopher Theophrastus described many types in his whopping *Enquiry into Plants*. The flowers made their way to Britain with the Romans, who believed the sap contained healing powers, but it actually contains irritant crystals.

I followed a footpath lined with yellow daffodils into a green field of winter wheat seedlings, busy with fieldfares. I got a buzz as a buzzard launched from a hedge and flew into the woods. Its mate took flight from the ground, swerving its bulk into the trees as best it could, for they're built for power, not poise.

Buzzards have a four-foot wingspan and weigh up to a kilogram, with fearsome hooked beak and sharp talons. They are plentiful these days, but almost disappeared from our skies in the 1950s following persecution by gamekeepers and pesticide use.

Male buzzards fly acrobatic displays to impress potential mates. Their trademark rollercoaster move involves flying high then plummeting, twisting and turning as they fall. Shifting baselines have been a regular and depressing feature of this project, but there are occasional positive examples too. I still get excited when I see a buzzard soaring overhead, even though they are common today. The population has risen fourfold in my lifetime since they received legal protection. We need more baselines heading in the right direction like this.

A golf course took up a large proportion of today's square. Pleasant enough, if somewhat artificial and dull. The landscape I mean, not the game (although, now I come to think about it…). I strolled down a footpath that ran alongside the 16th fairway, glancing only idly at an oak tree speckled with lichen. But then I stopped to use the Seek app to identify what I was looking at on the trunk.

The app identified rough speckled shield lichen, bushy cartilage lichen, monk's hood lichen and yellow shore lichen. Soon I was drawn in, noticing more and more, so many colours and textures, and falling down the tunnel of fascination. And there I had been thinking this golf course was a bit boring and empty. There was so much to see on this one tree that I'd almost dismissed with a single glance.

The more I pay attention, the more I notice. The more I notice, the more I learn. The more I learn, the more I enjoy. The more I enjoy, the more I pay attention. This positive feedback loop of learning and loving is what schools dream of generating in the classroom, but in my case, it didn't take hold until I began mooching around my neighbourhood with a camera, a notebook and a handful of apps.

On my map there are no forests free from litter, no hills to raise the heartbeat, no clean rivers to swim in, no ocean with crashing waves, no town bustling with people who enjoy the same things as I do. But the days when I set out to explore – to cycle, walk, photograph, sit, and think – were becoming the highlight of my weeks. Out here I did not notice what I was missing, but rather celebrated all that I was finding.

MARCH

RENEWAL

'I hope you love birds too. It is economical. It saves going to heaven.'

Emily Dickinson

I was drawn by the distinct scent of freshwater. It's such a fine, uplifting odour. 'Long enough in the desert a man, like other animals, can learn to smell water,' wrote the late Edward Abbey, American author and environmental activist, in *Desert Solitaire*. Far from a desert, across the railway tracks behind an industrial park, I found a misty, moody, monochrome fishing lake lined with rushes.

A heron circled overhead, stately and assured. Black-and-white tufted ducks careened in, to land with a waterski skid. Coots drifted over the smooth surface. They always draw a wry reminiscence from me because a thousand lifetimes ago I studied the 'agonistic communication' of coots for my university dissertation, whatever that means. But my heart had already clocked off from academia by then and I was getting ready to hit the road. I had requested to do my fieldwork in Africa, sniffing the opportunity to wangle an adventure out of a degree. But the professor was wise to my scheming and I was unceremoniously packed off to sit by a chilly Edinburgh duckpond for weeks, much like the one I'd discovered today. I smiled at my youthful disappointment and turned away from the coots.

Beyond the lake lay a river. From studying my map, I had hoped it might be the first swimmable span of water I'd found so far. One glance disabused me of that idea. The banks were sheer, dropping ten feet or more to low-tide mudbanks and a swift current. No good for swimming, perhaps, but appealing to wildlife. Herons chattered and squabbled in a treetop heronry on the bank. They gather for three months in the spring to breed, and sometimes their large communes of nests are reused over many generations. Another heron was poised like a single, silent sentinel on the riverbank, with its bayonet beak and stoically hunched shoulders atop knobbly stilt legs.

Farther along the riverbank stood a hexagonal pillbox from the Second World War, its thick concrete walls now wreathed in ivy and vines. It was still standing strong in the fog. Or, I asked myself, was it mist? The difference, it turns out, is only in the eye of the beholder, for mist and clouds and fog are essentially the same thing, with the differences depending only upon where you are looking from. Mist and fog form when saturated air condenses into droplets that hang suspended in the air. Saturation happens more easily when temperatures are low, so the phenomenon is more likely when the air is both humid and cold. If you look up at a mountain top it may seem to be shrouded in clouds. But hike up the mountainside and you'll find yourself entering a light mist. Keep on climbing into the heart of it and at some point you'll lose visibility in the fog. Such ambiguity felt like excellent advice. I could see the industrial estate I was heading to as a trudge

around warehouses and Portakabins, or I could consider it an adventure exploring somewhere I had never been before.

I crossed impersonal expanses of concrete, past litter, razor fencing and men in overalls with Eastern European accents. Yellow colt's foot flowers grew in the gutters among the padlocked shipping containers and tattered office chairs positioned outside fire doors for cheeky fag breaks. I saw a shopping trolley upended in a stagnant pool of water rainbowed with oil, and another one wedged overhead into the legs of a pylon.

A fragment of public footpath survived, hemmed in by chain-link fencing and lined with brambles and beer cans. Someone had garlanded a scraggy hawthorn tree with strings of seashells, a rubber duck, the head of an Action Man figure, and a traffic cone. I followed the path behind the warehouses, then crossed a railway line on one of those unguarded level crossings you occasionally find in quiet places. Other than a sign warning you to be careful, you're very much on your own. I like them as they feel exposed and trust you to make your own judgements.

The eastern half of today's grid square was a deserted area of blonde reedbeds. These are transitional habitats between water and land, waterlogged yet covered in vegetation and home to such elusive bird species as the bearded tit and the bittern. Plants have adapted to this wet habitat with modified stems, called rhizomes, that spread horizontally and shoot up into tall reeds that help aerate the underwater roots.

I thought about the reedbeds in terms such as 'peaceful' and 'uplifting'; very dissimilar in tone from the industrial yards a few hundred metres away. The way environments affect your mood like this is linked to the notion of nature connectedness and our relationship with the natural world. By tuning in to nature, and then enjoying and protecting it, nature connectedness helps individuals and society to build healthier and more sustainable relationships with the natural world.

When Miles Richardson, Professor of Human Factors and Nature Connectedness, began a research blog, his first mission was to show that noting three good things in nature every day for a week could lead to a long-term increase in how connected you feel to nature. The same principle applies to the Wildlife Trust's annual 30 Days Wild challenge. I felt confident that concentrating hard this year was helping me to forge some lasting habits. Research has shown that it is how much you concentrate when outdoors that affects your well-being, more than just the amount of time you spend standing out in the rain.

It is common when considering the challenges of nature connectedness to bemoan our addiction to screens and the digital world. But I prefer to embrace technology and celebrate the challenge of geocaching treasure hunts, games such as Pokémon Go, and apps such as 1000 Hours Outside, for encouraging kids to get out and run through the woods. The various apps I was using to learn more about nature and my landscape this year were invaluable (see page 364 for suggestions), despite the irony of needing a phone to help me observe the world around me.

Today's grid square was memorable for containing a rare error on my usually flawless Ordnance Survey map. A convenient footbridge across the river turned out to be some sort of chemical pipe. It was wrapped in barbed wire so I couldn't just trespass my way across it. I had to cycle downstream for several miles to the nearest bridge. Arriving on the far bank of the river, at last, I found a modern sort of farm: a solar farm.

The gleaming expanse of blue-black silicon panels was spread over a few hundred square metres, facing south and angled optimistically towards a sun still well hidden by mist. Solar energy is important for addressing the climate crisis as it is limitless, free and clean. Cover just a quarter of the Sahara desert in solar panels, hypothetically, and you'd meet the entire world's energy needs.

Although the first solar panel was invented in 1883, technology capable of harnessing significant amounts of energy only took off from the 1950s. Until recently, it was too expensive for mainstream adoption. But the cost per watt has plummeted since 2010, and solar energy now plays a prominent role in the world's transition from fossil fuels.

Away from the solar farm, bushes and saplings were rewilding an area that had once been covered by a Victorian brick factory. Young plants were bursting into life wherever I looked. Two centuries ago, this spot had been filled with railway lines, kilns, brick-drying sheds, warehouses, wharves, vast walls of stone, and chimneys belching smoke. It would have been a whirlwind of industry, noise and money-making. Yet almost nothing remained. That thriving enterprise was long gone, like an English Ozymandias. Nature had returned in its place, making this square feel like the wildest I had visited so far.

Among all the things I've worried about in the dark first half of this year, a beam of sunshine that warmed my heart was how reliably nature bounces back. In the gloom of winter, I had perhaps half wanted to be disillusioned. Maybe it was my pessimism at consigning myself to twelve months on a small map I didn't much like, but I had actually really enjoyed the wildness of some of the places I have found, and now spring was infusing me with its exuberant enthusiasm. That the wildness was all reclaimed from industry cheered me even more.

If we don't wreck things too much, habitats do recover, and we can make environments wild and wonderful again. That's no consolation, sadly, for the thousands of species disappearing annually, thousands of times faster than the natural extinction rate. Experts argue over how many species there actually are, and whether 0.01 or 0.1 percent of them go extinct annually. Either way, we are destroying between 10,000 and 100,000 species every year.

While 99 percent of all organisms that have ever lived have already died out, the fact that we are responsible, right now, for the sixth mass extinction event in history means I don't feel mollified that it is part of the natural order of things. I am ashamed that more than half of British species have declined in my lifetime. We have lost so much. Although, as naturalist Chris Packham pointed out in his passionate People's Manifesto for Wildlife, 'lost means inadvertently misplaced.

No, our wildlife has been killed, starved, poisoned, ploughed up or concreted over.' This happens not because we are evil, but because we look on and do nothing.

But still, despite everything, nature's potential for recovery is tremendous once humans get out of the way. For example, during the Covid lockdowns, civets returned to once-busy streets in India, deer roamed the roads in Japan and monkeys enjoyed a splash in a swimming pool in Mumbai. The German Green Belt is a nature reserve running the length of the country down the old fortified Iron Curtain. And Chernobyl, off-limits to humans for decades, has become a haven for wildlife, with lynx, bison and deer roaming through regenerated forests.

Ecosystems are at their most successful and stable when there are populations of large, native keystone species such as these holding everything together.* Bring back these ecological engineers and degraded habitats spring back to life through a trophic cascade effect. Reintroductions of keystone species can be controversial, though, such as the bison in Blean Woods near Canterbury and beavers across Britain. The anxiety is another example of shifting baseline syndrome, as we think that the nature of our childhood was 'normal' without these species, despite evidence that animals such as beavers have so many positive effects, such as improving tree growth, increasing biodiversity, and stabilising riverbanks.**

If reintroducing species to restore habitats feels a bit radical, I also like the simple notion of doing nothing for nature. Just stepping back and leaving wild spaces in gardens, verges and unneeded farmland allows nature to return. This is what the Knepp Estate in West Sussex did on its way to becoming England's most famous rewilding project, allowing 3,500 acres of unprofitable farmland to return to the wild.

**Keystone species are not always large mammals. Sturgeon, dung beetles and bog moss all make a tremendous impact on their habitats, and spawning salmon nourish trees on riverbanks as their decaying bodies provide lots of nitrogen.*

***In* Three Against the Wilderness, *Eric Collier recounts how an elderly First Nations lady taught him that returning beavers to his valley in British Colombia would lead to a total recovery of the area. 'Until white man come, Indian just kill beaver now an' then s'pose he want meat, or skin for blanket,' she said. 'And then, always the creek is full of beaver… S'pose once again the creek full of beavers, maybe trout come back. And ducks and geese come back too, and big marshes be full of muskrats again all same when me little girl.'*

Knepp was a failing farm, stripped of biodiversity, inefficient and struggling. Today it is home to populations of rare nightingales, purple emperor butterflies and turtle doves, as well as being a thriving success story for nature tourism. The estate has wildlife safaris and places to stay, and sells wild range meat from the free-roaming herds of Old English longhorn cattle, Tamworth pigs, and red and fallow deer whose browsing behaviour helps create new habitats for wildlife. Seeing before and after photos of rewilded sites such as Knepp, or Carrifran in the Borders, Dundreggan in the Highlands and Ennerdale in Cumbria, always fills me with hope.

I pushed through undergrowth to reach a reservoir that had been fenced off by the water company. It seemed I was not the only person sorry to be denied access, for a fence panel had been dislodged and a path led down through the trees to the shore.

If you ignored the bottles of cheap imported vodka and discarded fishing tackle, then it was a beautiful spot, especially now the sun had come out. The still water glimmered and reflected the blue sky. I had been practising this dichotomy of control mindset a lot during my wanderings across the map, trying to have the serenity to accept the things I could not change, the courage to change the things I could, and the wisdom to know the difference.

In this case, it involved me ignoring the litter, enjoying the view and appreciating having a lake to myself on a sunny day, with clean water and the merry sound of birdsong. I stripped off and waded in for my first wild swim of the year. It was in equal parts nippy and invigorating. When I departed, I filled my rucksack with litter from the lakeshore to leave a positive impact on where I had been, doing my bit for the cause of opening up the countryside and treating it responsibly.

There might be little wilderness on this map of mine, but there was plenty of wildness if you looked for it, a messy, energising sense of the living world being all around us and pushing hard to make itself felt.

BLOSSOM

'I feel as if I had got a new sense, or rather I realise what was incredible to me before, that there is a new life in nature beginning to awake… It is whispered through all the aisles of the forest that another spring is approaching. The wood mouse listens at the mouth of his burrow, and the chickadee passes the news along.'

Henry David Thoreau

'March, month of "many weathers",' grumbled John Clare, the peasant poet, and I thought of him as I sheltered from a shower beneath a church's lychgate. Lych is derived from the Old English word *lich*, meaning corpse, and the lychgate was where a group bringing a body for burial would meet the priest. These were the mad March days of rain then sun then wind then rain then breakfast. I spent all of yesterday's sunshine indoors, filling in tedious tax stuff for my accountant and looking forward to today's outing. It was foolish to have expected that the weather today would still be mild.

We are told to expect March winds in the proverb that 'March comes in like a lion and goes out like a lamb'. Germans used to welcome the month's thunderstorms, though, believing they led to a bountiful harvest. 'When March blows its horn, your barn will be filled with hay and corn.'

Once the shower passed, I wandered round the churchyard for a while, looking at gravestones. I preferred those with biographical details about the life lived in the dash between the birth and death dates, rather than just banal tropes.

'For many years, a well-respected local veterinary surgeon.'

'Loving mum, nurse, entrepreneur, publican.'

'A generous and respected friend.'

One tomb quoted William Blake's words that are a fine guide to exploring a single map:

'To see a world in a grain of sand
And a heaven in a wild flower,
Hold infinity in the palm of your hand
And eternity in an hour.'

And there was a cautionary admonishment from 1786 to *carpe* my *diems*:

'Stop traveller, and cast an eye,
As you are now, so once was I.
As I am now, so must you be,
Therefore prepare to follow me.'

A fresh grave bore a wreath from the Kano State Government. Intrigued, I googled the name of the lady buried here in blustery England, far from the hot millet and sorghum fields of northern Nigeria. But all I could find about her life was a link to the expired video stream of her socially distanced online funeral during the Covid lockdown.

That global drama was very different from the one that claimed the life of George, a pilot born and buried in this hamlet. He died in the Second World War, aged eighteen, in an accidental aerial collision with a pilot from the spectacular Deep Cove inlet in distant British Columbia. George's grave was marked with a red ceramic poppy, alongside his brother, Alfred, who was killed in Germany in the final few months of the conflict. I thought of the reverberations of that war rumbling through communities around the globe as I stood before their father's grave, who outlived both his sons by many years.

The church lay in the foot of a valley, so I dropped into a low gear to cycle uphill through the rain, past a bungalow called Sunset Towers with a pristine white picket fence and privet hedge. A footpath led into a wood, and I sat down to drink my coffee on a log in an airy clearing where an old beech tree had fallen. The tree's root system, now pointing skyward, was three metres across, yet less than a metre deep. It often surprises me how shallow are the roots of trees, although I suppose I only ever see those that have fallen over and failed at their job. Trees rarely root more than two metres deep, and most roots lie within the top sixty centimetres of earth.

I was sharing my log (but not my coffee) with a gloriously weird, slimy pink fungus called a wrinkled peach, fond of rotting elm, though tolerant of other hardwoods. With the sad demise of our elms from Dutch elm disease, these fungi are now rare in Britain. The infection killed millions of the trees when it was imported from Canada in the 1960s and changed our landscape for ever. The disease, spread by the elm bark beetle, is still moving north today.

History is repeating itself with ash dieback, a new fungal infection spreading across Britain. It arrived via imported saplings, and highlights the hazard of the international plant trade. The disease spreads via spores that blow tens of miles on the breeze and may well kill 80 percent of our 180 million ash trees.

The footpath out of the wood was lined with shiny, spiky holly trees studded with red berries. Early green shoots of bluebell leaves sprouted under the delicate white and pink blossoms of blackthorn bushes and cherry trees. In Japan, cherry blossom, *sakura*, reminds people of renewal. They celebrate with picnics beneath the flowering trees, a custom known as *hanami*. The blossom epitomises the concept of *wabi-sabi*, accepting the beauty of imperfections and the ephemerality of everything.

Petals drifted in the breeze like snow as I emerged from the wood, pushing through hawthorn trees just beginning to bud, and out onto an expanse of playing fields. A buckled and smashed tennis racket on the ground painted a succinct picture of the frustration of defeat. The wind moaned through the skeletal tree branches, chilling me now that I was away from the wood's shelter. The sports pitches were deserted and the cricket sightscreens were all huddled together in one corner, waiting for the joy of summer to return.

I climbed a stile into farmland. A shiny blue Ford Fiesta was parked in the corner of a field. I wheeled my bike round fields of grazed grass,

and through some that were scrubby, brown and full of thistles. Most contained dazzling green winter wheat seedlings, planted either as a cash crop for next summer or as a cover crop to protect the soil from water run-off, weeds, pests and diseases. Cover crops are also helpful for fixing nitrogen and for wildlife. Bare earth is terrible for erosion, nutrient loss and carbon release.

I dropped back into the valley that ran the length of today's square, then pedalled up the other side, lamenting my lack of fitness as I huffed and puffed in bottom gear. I was riding so slowly that I had time to watch a plump bumblebee busying itself around a hole by the road.

'That's early,' I thought.

I was surprised to find that I was correct, for once. The early bumblebee (of which I'd never heard) is one of our prettiest and smallest bumblebees, commonly seen in early spring, hence its name. It has an orange tail and pale-yellow stripes. Early bumblebees nest underground or in the abandoned nests of mammals or birds, gathering in small colonies with a queen, who emerges between March and May.

I kept battling intrepidly and ineptly up that little hill, in order to reach the first trig point I'd encountered on my map. Having not been up a mountain all year, it felt like meeting an old friend again as I patted the concrete block, a whopping 110 metres above sea level in the middle of a muddy field. It was a contrasting world from camping a couple of years ago by the trig point on top of Suilven in Scotland, my favourite mountain, with staggering views in all directions of peaks, lochs and islands.

Trig points were originally used for the re-triangulation of Britain in 1936, beginning somewhere in a field in Northamptonshire and branching out from there across the country. They enabled the precise measurement of angles that improved the accuracy of Ordnance Survey maps over the next twenty-six years. Mapping technology has moved on since then and the country's 6,000 trig points stand now as essentially obsolete novelties.

The prolific hillwalker Rob Woodall became the first person to visit every trig point, on a mission that took him from Lizard Point in Cornwall to Unst in the Shetlands. He climbed Ben Nevis, Britain's highest peak, and bagged the trig point by the Little Ouse in Norfolk, which lies 1,346 metres lower – one metre below sea level, in fact. It took two years of campaigning before he was granted permission to visit the privately owned trig point in Fort Borstal, Kent. At the end of his journey he said, 'I'm not sure what I will do next but I'm not sure I could ever take up stamp collecting.'

I pulled up my hood against the icy wind and looked around as I ate my salt and vinegar crisps. Fields and woods fell away towards the red-white streaks of motorway lights. It was a pleasant view, even on this murky day. I could see all of today's square, and also landmarks in several others that I had visited, such as the giant pylon and the viaduct. It was so satisfying to be gradually slotting the puzzle pieces of my map together and bringing it all to life.

PIGS

'Early one morning, any morning, we can set out, with the least possible baggage, and discover the world.

Thomas A. Clark

I locked my bike by the pond on the village green. It was a quiet morning and nobody was about. Village greens conjure peaceful images of cricket matches, community celebrations and maypole dances. But historically, village greens were about more than recreation. Since the Middle Ages they have been an area of common grassland for the use of everyone, often with a pond where fish were reared, cartwheels soaked to prevent them shrinking, clothes washed, cattle watered, and dishonest traders punished on ducking stools as social humiliation.

Completing today's bucolic scene was an old flint-and-brick oast house. Buildings like these were once used to dry hops for brewing beer, so the distinctive conical shape is common in hop-growing areas. I set off along a narrow lane beneath an archway of hedges and trees. A notice pinned to a fence said 'Do not feed horses no carrot or apple.' Horses' hooves had chewed the earth to sloppy mud, so I picked my way carefully down the edge.

A red sign declaring 'PRIVATE GROUNDS' was nailed to an old

beech tree on the edge of a copse. 'NO THRU ACCESS' read another. 'PRIAVATE [sic]. NO PARKING. RESIDENTS ONLY' warned a third. Even where there were footpaths, it felt as though they'd been allowed only grudgingly, with fences and cautionary signs keeping me strictly on the narrowest strip of land it was possible to walk on. It was a cheerless affair, a mean-spirited granting of minimal space. At one point the path became a claustrophobic tunnel between high fence panels that was barely wide enough for my shoulders.

A squirrel sitting on this fence spotted me. Its nearest safe tree was about twenty metres away. Unfortunately for the furry fellow, I was between the tree and him. The squirrel decided that the lure of the tree was greater than the terror of the human, so it sprinted along the top of the fence, a clitter clatter of tiny claws, zooming straight past me within easy arm's reach, before leaping for the safety of the high branches.

As a fond feeder of birds, I am in constant conflict with the grey squirrels in my garden who try to scoff all the seed I put out. I have a squirrel-proof bird feeder outside my writing shed, complete with a squirrel-proof concave baffle to stop them climbing the pole. But even so, the blighters drop from trees or leap from my shed roof to try to get at the food. I have to admire their agility, even though their greed and population size annoy me.

Squirrels can make enormous vertical or horizontal leaps using their powerful double-jointed legs. They can swivel their back ankle

joints and so run down a tree as easily as they run up it. Cats don't have this hypermobility, hence their tendency to get stuck up trees.

Feeding birds is a simple pleasure and an easy way to connect to nature. There is a charming line in Ronan Hession's novel *Leonard and Hungry Paul* about a mother who 'looked after everyone in her life as though they were her garden birds: that is to say, with unconditional pleasure and generosity'.

But, like many things, feeding birds also has a negative environmental side. My heart sank a little when I learnt this. It is so hard to cause no damage even when you're trying to do the right thing, such as switching from cows' milk, which is terrible for emissions and polluting rivers, then learning that your smug almond milk is bad for bees, or flipping from burgers' awful impact on deforestation to discovering that each smashed avocado you post on Instagram guzzles 320 litres of precious water.

Feeding birds can spread disease and it skews populations towards those that most benefit from the calorie boost: adaptable and aggressive generalists such as great tits, nuthatches and parakeets. Timid birds, such as the wood warbler and marsh tit, have suffered sharp declines as a result.

This grid square felt like an old-fashioned rural landscape. There were farms with cows, sheep, pigs, horses, goats and chickens, and none of the old McDonald's litter I'd grown used to. The rolling countryside was thick with woodland and there was also more birdsong than a week ago. More green shoots too, as leaf buds sprayed the woods in their first hopeful emerald mist.

I nibbled some fresh hawthorn leaves, a tasty morsel once known as 'bread and cheese', which made me think of the yellowhammer I'd seen earlier, a bird whose song sounds like it is calling 'a little bit of bread and no cheese' over and over. There was an old folk anecdote that hawthorn blossom reeked like London's great plague. Bodies used to lie in the house for days before burial, so people were familiar with the smell of death. Scientists have since discovered that the blossom does actually contain trimethylamine, a chemical present in decaying flesh.

Three buzzards circled high overhead, and a mischief of concerned magpies grew agitated by their presence. I began counting the magpies, reciting the rhyme I remembered from childhood.

'One for sorrow,
Two for mirth,
Three for a funeral,
And four for a birth.'

Magpies have been linked to superstitions and omens for centuries, with the earliest variation of the famous rhyme first written down in 1780.

Folklore surrounds the magpie. Salute it to stave off misfortune, greet it for luck, or fear it being in collusion with the Devil. For this is the bird, it was said, that refused to mourn Christ's death with the other birds, that refused to enter Noah's ark, and instead alighted on the roof and cursed throughout the voyage. In Korea, by contrast, they say that a magpie predicts the auspicious arrival of visitors.

The magpie's reputation for thievery stems back to 1815, with a French melodrama featuring a servant girl sentenced to death for stealing silverware, though the thief turned out to be the master's pet magpie. Rossini then turned the story into the opera *The Thieving Magpie*, besmirching the poor bird's name for ever. Scientists have found no evidence that they're more attracted to shiny objects than other birds.

And what of magpies' reputation for killing birds? They are noisy and vigorous predators, certainly, but there is little evidence they have an adverse effect on songbird populations. Ecologist Tim Birkhead defends them, saying, 'cats are undoubtedly a monumental threat to songbirds, but it's magpies that incur the wrath of the average bird lover. If magpies were rare, people would travel far to see them. In bright sunlight they are the most exquisitely beautiful birds, with that lovely long tail and iridescence.'

I walked over some more fields and encountered my map's first pigs, fenced into an area of woodland that they had turned completely to mud. The pigs looked as happy as... well, I can't think of a simile, but they looked very content.

'Morning, pigs,' I called.

Three little ones paused their rooting to stare at me from beneath floppy ears. Pigs have been put out to pasture in woodland for centuries. Allowing them to forage for acorns, beech mast or chestnuts, a practice known as pannage, began in medieval times as a privilege granted on common land or in royal forests. As well as fattening up the pigs with free food, pannage was a useful way of turning the soil and also protected ponies and cattle from hoovering up too many green acorns which are poisonous to them.

So long as the pigs are rotated through different areas in small numbers, this ancient system of silvopasture or agroforestry is sustainable. A mixture of trees, pasture and a few grazing animals is better than grassland alone for sequestering carbon and helping to counteract the livestock's emissions.

Besides helping the environment, agroforestry generates income on separate levels, from foraged nuts, berries and fungi, to rearing livestock and timber production. This shields farmers from market fluctuations. Agroforestry is good for the farm, good for the farmer, and can even be good for the planet if done correctly.

Leaving the pigs, I walked past patches of primroses, a plump stuffed panda perched in the fork of a pollarded tree, and a ramshackle wooden hut. If the homemade cabin was a kids' playground, the construction was impressive. If adults built it for some other function, then it was very much beyond its prime.

I often come across mysteries in the woods like this, signs that someone once did something here, for some reason, sometime. I wish I could meet everyone who has ever walked past one spot, all gathered here at the same time, from dog walkers in Gore-Tex, back through woodsmen and deserting soldiers, hunting lords and peasant poachers,

back further to legionnaires and Neanderthals, and then all the way back through the mammoths to the iguanodons.

Across the valley, an old timber framed cottage was tucked onto a hillside in a fold of woodland, like a scene from *Cider with Rosie*. I headed towards it. A sign cautioned drivers to be careful of badgers crossing the narrow lane. Lining the grassy public footpath leading to the house were eight small, homemade wooden crosses, the surrounding earth sprinkled with yellow primroses, and each cross bearing the dates and name of a beloved pet.

I stopped to chat over the low fence with an elderly lady who was pottering around her garden. She told me cheerfully that her home had been built in 1900 and 'I've been here nearly as long as that.'

I asked what lived in her pond, for I have been on a long and fruitless mission to find frogspawn on my map.

'Oh yes,' she told me. 'We get loads of frogs. Newts too. And dragonflies.'

As a boy, I used to push through the beech hedge at the bottom of our garden, out into the fields, and collect gloopy balls of frogspawn from a nearby spring. We kept them in an old sink outside the back door until they hatched and hopped away. It was fascinating to watch the tadpoles grow into froglets. I'd like to watch the same miracle unfolding today outside my shed. But because she was a frail old lady, I didn't want to trouble her by asking if I could come into her garden and collect a bit of her frogspawn. She would probably have thought me very weird had I filled up my water bottle with frogspawn anyway. I shall continue my search.

For meteorologists, spring begins at the start of March. For astronomers, it is the equinox. For me, the season was already well underway with the first frogspawn and the beginning of a new generation of wildlife.

HOUSES

'Is it so small a thing
To have enjoy'd the sun,
To have lived light in the spring,
To have loved, to have thought, to have done.

Matthew Arnold

Blackthorn blossom decorated every lane this week. It was late March and the best time to spot the difference between hawthorn and blackthorn. Blackthorn trees blossom before their leaves appear, while hawthorn does it the other way round. We use many cues to connect what we see with the seasons (fairy lights at Christmas, for example), and making a conscious effort to be observant each week was building a richer natural calendar in my mind than I'd ever had before. I hope next year I will instinctively think, 'Blossom season, and that hedgerow is blackthorn, not hawthorn. It must be late March.'

Now I heard the year's first chiffchaff chirping away, a call like a tiny blacksmith hammering an anvil all day long, 'chiff-chaff-chiff-chaff'. It is easily confused with the great tit's '*teach*-er *teach*-er' chirp. Not many people get excited by a little brown bird with a monotonous song. But I enjoyed celebrating a feisty six-gram bird that had flown all the way here from Africa. I was becoming aware of so many things that had passed me by in all my decades alive. The sense of amazement was boosted by small new abilities such as distinguishing a chiffchaff from a great tit by their songs.

I arrived in today's grid square down a busy road, cars swooping back and forth, that demanded all my concentration. A workman battered the pavement with a pneumatic drill, and I had to turn off the road onto a quiet street of new houses before I could quieten my mind and settle into the slow rhythm of exploring.

Just a stone's thrown from the railway station, this cul-de-sac was prime real estate for wealthy people commuting into the city. The homes were huge, with oversized cars parked outside. But all this bigness came at the expense of any outdoor space. They had squeezed ten identical buildings onto a plot of land that would have been the size of one garden for a home like this in earlier times. This tug between houses and space was to be a recurring theme on today's ride.

I waited for an old lady to pass on her squeaky mobility scooter, then rode behind the station into a neighbourhood of circular residential roads, lined with 1960s bungalows set back in tidy gardens. One was called Crianlarich, with a garden filled with purple heather. I wondered what had brought the owners south from the scenic gateway to the Highlands.

Bungalows feel quintessentially British, like cups of tea or the Royal Family. I live in one myself. Yet none of those things is native. As with much in British life, the humble bungalow is wrapped in the days of Empire. The word bungalow comes from the Hindi *bangla*, meaning 'belonging to Bengal'. It was used to describe colonial cottages in India. I remembered seeing some fine examples when walking through Kerala's forested hills. Settlers returned to Britain with enthusiastic memories of the Raj lifestyle and the charming homes they had lived in (and probably a fondness for tea and royalty too).

The next circle of houses wasn't as affluent as the new-builds by the station, nor retirement homes like the bungalows. This was the domain of the mid-tier commuter, successful but still striving, although the nearby train line into the city undoubtedly made the rows of semi-detached mock-Tudor homes eye-wateringly expensive. The streets were perfectly fine and pleasant if you're after a fine and pleasant life, but I

yearned to be back among mountains, ocean waves, rushing rivers, and the sort of people who gravitate to those places.

Yet I also understood that there are a thousand and one reasons why we can't all live in log cabins in Alaska, not even that small minority of the population who would actually want to. For those of us in that category, yearning to be free, appreciating the wildness that is around us is a better approach than lamenting what is not.

Here in town, it was bin day, so there was a full wheelie bin or two lined up outside every home. We have an efficient system that whisks away our rubbish and recycling, but perhaps it is too efficient, because once it's out of sight, we forget all about its impact. British households produce 27 million tonnes of waste a year, more than 400kg per person. A significant amount of that is spoilt food, as many families throw away up to 20 percent of their food, worth about £800 per year. If global food waste was a country, it would be the third-largest greenhouse gas emitter. Every year, the world chucks out two billion tonnes of rubbish. Load that onto lorries and they'd lap the world twenty-four times.*

'Enough of houses and landfill and concrete streets!' I cried to myself, and pedalled on in search of something wild to soothe my soul.

**To highlight our wastefulness, activist Rob Greenfield lived like the average American for a month, but discarded nothing. Using a custom-made suit, he carried every piece of rubbish he generated everywhere he want. By the end of the month in New York, Rob was squeezing through doorways, struggling to pass subway barriers, and prompting conversation wherever he went.*

I found it, or the beginnings of it, in a park covered in trees, grass and transitional scrub. Forty years ago, the area had been farmland, but it was now covered with an impregnable mesh of saplings and bushes, all self-sown and being allowed to develop naturally. Footpaths criss-crossed through the dense cover. I sat on a bench in a clearing and took stock of things. The motorway still roared beyond the wood, but it was a low, steady noise that I could tune out or pretend it was a waterfall. I looked around at the trees and blossom, listened to the birds, and remembered that it was the week of the spring equinox, when the hours of light and darkness are equal. Summer was on its way.

For ancient people, the spring equinox was one of the year's most important markers. Everyone would have been tuned in to the shifting of the seasons. Relatively speaking, we have only recently become so distanced from the natural world to live in indoor bubbles with regulated light and temperature. Communities used to celebrate the equinox with bonfires as they feasted on the last of the winter stores. It falls in the middle of a period known as 'the hungry gap', when little fresh produce is available, so a festival helped to raise famished spirits.

This itch to welcome the spring became the pagan festival of Ostara, personified by the goddess of dawn and rebirth, a young woman surrounded by light, flowers and fresh growth. It celebrated lighter days, new life, and renewed hope. Ostara's name came from the old German word for east, where the sun rises. Many of these rituals remain today at Easter, passed down to us through Celts, Saxons and Romans, such as colourful eggs, rabbits and baskets of sweets.

The northern half of today's square was striped with tight contour lines and divided between a wood and a housing development perched on the same hillside. Half the wood had been cleared to build the homes. I was interested to cycle around the houses on the bare hillside and then compare it with the woodland next door. We love woods yet need houses. But we also need woods and can love our homes. The conflict of demands on our land was very apparent on that single hillside.

The houses were, frankly, amazing. I had not seen so many lavish homes anywhere else on my map. There were footballer-style mansions behind high gates, painted clapboard houses such as you might find in an affluent American suburb, imitations of old rectories filled with Labradors and hockey sticks, and *Grand Designs* buildings that were all crisp, straight lines, massive plate-glass windows, and sans serif nameplates. They were fine homes, but I felt an unease as I pedalled along the quiet lane about the inherited inequalities on my map. The gulf in wealth between these residents and the poor estates of crowded tower blocks I had ridden past to get here keeps being passed down from generation to generation.

Why does our society obsess over measuring growth and wealth as the sole indicators of success? India has a booming economy measured in Gross Domestic Product, but has shameful inequality and is one of the world's most depressed countries. Bhutan, on the other hand, scores poorly for GDP, but ranks highly for Gross National Happiness, which it values highly. Where do we want our priorities to lie? How should we measure our societies?

Chasing growth, bigger houses, and ever more stuff to buy and bin is literally unsustainable on a planet of finite resources. As I cycled around, I recalled Kate Raworth's *Doughnut Economics* book. It emphasises the need to balance our obsession with growth against quality of life for everyone and the earth's environmental limits. Only in the middle of this 'doughnut' can we find the 'ecologically safe and socially just space' that society should aim for.

Amen, I thought. And went off to have fun riding my bike around the woods.

APRIL

VINEYARDS

'If ever I saw blessing in the air
I see it now in this still early day
Where lemon-green the vaporous morning drips
Wet sunlight on the powder of my eye.'

Laurie Lee

Out into the delirium of spring, riding fast and light-hearted towards today's grid square. Birds belting out love songs in every hedgerow. The first blush of sunshine in the oilseed rape fields, pretty but terrible for leaching nitrates into waterways. The first sulphurous brimstone butterfly, a yellow that put the 'butter' into butterfly. In *Every Day Nature,* Andy Beer suggests you note the first date you spot one and call it your Brimstone Day each year. I liked that idea.

My computer calendar already reminds me of the various dates of the first snowdrop and daffodil outside my shed in recent years. The first green leaf on the tree by my window, the return of goldfinches to my feeder, the first swift and, from today, the first brimstone butterfly. I enjoy seeing these differences in nature's calendar year on year, the phenology of where I live.

It is a start, but my novice observations are a long way from those of the Reverend Gilbert White, whose detailed decades of notes about the natural world around his village resulted in *The Natural History of Selborne*, a book that has remained continuously in print since 1789. He was a pioneering and inquisitive natural historian with astonishing powers of observation. His writing also offers an invaluable insight into rural life in the 18th century. It was often carried by emigrants to North America and Australia who wanted a nostalgic reminder of home.

White paid minute attention to nature and recorded it diligently, a practice he called 'observing narrowly'. The more he focused, the more engrossed he became in the small wonders on his doorstep in Hampshire. For example, he observed that owls hoot in the note of B flat and surmised that willow wrens were actually three separate species by tiny differences in their songs and plumage: chiffchaff, willow warbler and wood warbler. I'm quite proud of myself if I even glimpse one of those as they dash from bush to bush, never mind playing spot the difference between them.

I locked my bike in a churchyard then walked past a couple of wooden thatched barns and a Victorian postbox wreathed in ivy. A plaque on the village hall commemorated the date when this valley had been saved from being commandeered by the Ministry of Defence for mine laying and infantry training. I crossed the cricket pitch (oh hurry along, summer!), passed a 'hedgehog crossing' sign and began climbing a path up the side of the valley.

No landscape in this country is fixed nor exists in its original state. The countryside is always in flux. And this valley had changed tremendously in recent years, with thousands of grapevines being planted across hundreds of acres. Producing wine in England on this scale is a mark of our evolving climate, and agricultural changes such as this look set to continue. In every direction, the land was now covered with rows of small vines, each staked out and protected from rabbits in a green sheath (hopefully biodegradable).

I paused on a bench at the head of the steep valley to catch my breath and look down over the vines towards the cricket pitch. Overhead, I heard one of my favourite birdsongs, the skylark's, which inspired George Meredith's poem 'The Lark Ascending', to which Ralph Vaughan Williams's musical composition was a response. Skylarks are ordinary-looking brown birds with a funky tuft on their

head. They nest on the ground in open areas across Eurasia and North Africa. What makes them special is the male's incredible singing throughout the summer months.

Skylarks burst up to great heights, almost 300 metres, hover, plummet, and then climb again – all while singing a beautiful, burbling song. It's like belting out arias while running uphill at top speed. They fly so high and sing so loudly that Tennyson described how, 'drowned in yonder living blue, the lark becomes a sightless song'.

New life was blooming across the land. Blue bird's eye speedwell flowers crept along the ground. White wood anemones sparkled in the shadows. Celandines beamed like little suns. Peacock butterflies fluttered their red wings in the sunshine, winking bold eyespots to mimic larger predators. One gigantic oak tree caught my eye when I remembered that its bare branches would soon be upholstered in green once again, but most trees were already pushing out pale leaves.

Even though it was now spring, this annual process of renewal started way back in the autumn, when trees respond to the shortening days, lengthening nights and lowering temperatures by beginning their dormancy. New buds have a 'cold requirement' so they don't grow too early, needing a minimum number of cold days followed by an adequate warm spell before bursting forth.

God, it was superb to be on the move! I'd donned shorts for the first time, the air smelt warm, the sun shone on my back, and the horizon of low hills broke up the monotony of the usual flat skyline. I too was coming back to life after a long hibernation. I loved it up there.

And there was so much to take in, everywhere. Here the first bluebell of the year, there a dunnock, a woodlouse, a budding beech tree, a nesting blackbird, a path I hadn't spotted, a pheasant's call, the first fragrance of wild garlic. All these things scrabbled through my mind all at once. It was overwhelming. So it was something of a relief when I found a mossy stump in the woods and decided to try sitting still for an hour to take everything in.

I felt I needed to do this to help immerse myself in this day and this grid square. It can be hard to do when you are busy. The less time we have, the more we probably need to do it. If you think time is racing too fast to waste sixty minutes sitting on a log, fear not: this hour will feel like an eternity. I set an alarm for the end, then put my phone and watch out of sight and out of reach. And then I just sat down on my stump and waited.

Richard Louv coined the phrase 'nature deficit disorder' in his influential book *Last Child in the Woods*. Our ruptured relationship with the natural world threatens our health, the economy and the future of our relationship with the planet. Louv highlighted its links to obesity, ADHD, isolation and depression, warning that 'time in nature is not leisure time; it's an essential investment in our children's health (and also, by the way, in our own)'.

Without spending time in nature, we lose awareness of a world that is real and vital, yet too often hidden on the far side of our windows, walls and screens. Nature, even a sliver, is available every day, everywhere, for everyone. It was up to me to observe it and appreciate it, alone on a stump in a wood.

'Alone is a fact,' wrote dancer and choreographer Twyla Tharp in *The Creative Habit*. 'Lonely is how you feel about that.'

Sitting in a wood for an hour with no pen or paper and nobody to talk to is like the methods of mindful meditation. It is both an invitation and a challenge. You are observing what is in your head, but not recording it. As your thoughts whirl, you can only notice them arrive and then allow them to leave. If you are fortunate, you might settle into a state described by the delightful Gaelic phrase *ciúnas gan uaigneas*, 'quietness without loneliness'.

My hour sprawled into a daunting expanse of time. I heard a dozen

bird calls and watched a bumblebee rustling in dead leaves. I wondered what was for lunch. I felt the sun shining on my face, 660 million tonnes of hydrogen burning every second, and that light and warmth already eight minutes old when it reached me. I closed my eyes and allowed myself to absorb it.

I was certain that the alarm clock had broken and that hours had passed. I squirmed. I yearned for the end. But, just as when I'd sat still for an hour back in January, my first thought upon finally hearing the bell was disappointment that the hour was over.

The biologist David Haskell is an expert at watching nature. He visited a one-square-metre patch of forest over the course of a year while writing his book *The Forest Unseen*. His rules to himself were simple: 'visit often, watch a year circle past, be quiet'.

One outcome of his observations was 'to realise that we create wonderful places by giving them our attention, not by finding "pristine" places that will bring wonder to us'.

It does not matter where you go. It matters only that you go.

DAISIES

'The year is but a succession of days, and I see that I could assign some office to each day which, summed up, would be the history of the year. Everything is done in season, and there is no time to spare.'

Henry David Thoreau

I passed a primary school in a forgotten-looking estate of identikit tower blocks as I cycled into today's grid square. The playground was full of joyous shrieks and laughter, and three colourful quotes were displayed on the wall:

'Somewhere, something incredible is waiting to be known' Carl Sagan.

'Education is not the filling of a pail, but the lighting of a fire' W.B. Yeats.

'The more you read, the more things that you will know. The more you learn, the more places you'll go' Dr Seuss.*

These are brilliant quotes for an education built on curiosity not box-ticking, but they also summed up what fascinated me about diving into my map. The diversity of living experiences across just twenty kilometres was stark, and I wasn't sure I belonged in any of them. I was an outsider in the towns I visited and felt more comfortable in the

**Dr Seuss, incidentally, was not a Doctor, but added it to his pen name to appease his father, who had wanted him to study medicine.*

countryside. Yet there, too, I am an imposter when writing about it. I don't work on the land or in conservation so don't really know what I'm talking about. I'm just a concerned onlooker at the car crash, hoping the professionals will arrive soon to save the day.

The experience of getting to know this area would be quite different if I felt attached to it, perhaps generationally hefted, or if I had a job that immersed me in the communities here, like our chatty postman who seems to know everyone and everywhere. Perhaps, I diagnosed myself nervously, my fondness for the open road and new horizons was actually a fear of commitment and a sense of isolation at not belonging anywhere?

Around the housing estate I saw kids bunking off school, mums with prams and cans of energy drink, and men in boxers smoking breakfast cigarettes on their doorsteps in the mid-morning sunshine. Meanwhile, a few hundred metres away, the smart cul-de-sacs with cherry blossom trees were deserted, with the kids all in school and the adults out at work. Seeing the variety of landscapes on my map was fun; the disparity of opportunities was more jarring.

These now-scruffy tower blocks were built in the housing crisis after the Second World War and opened by Prime Minister Clement Attlee. He declared, full of hope, 'we want people to have places they will love; places in which they will be happy, and where they will form a community, and have a social life and a civic life'.

I can't judge a place from walking around it for an hour, but the community library was 'temporarily closed until further notice', the shops sold little but junk food, and every small patch of grass bore signs warning 'No Ball Games. No Cycling.' On the plus side, there was a community gym with an outdoor weights area, a youth club with a football cage, and a café busy with old folk enjoying the sunshine and each other's company.

On one of the grass verges (where fun was forbidden), small mauve flowers grew among the dandelions and curls of dog shit. I learnt they were musk stork's-bill. What a name! What a delight! It was just an obscure little flower on a scrappy verge, but the name burnished it and made me smile. And the kids in the playground were still hollering. It's hard not to be optimistic about the world when wildflowers are blooming and children are laughing. I smiled and set off to see what else this

grid square might reveal.

I rode up and down straight Victorian terraces for a while, brick-built with four chimneys on each roof from the days of coal fires. Those polluting fires are long gone and the air is cleaner now. Swap electric bikes for some of the parked cars that now hem the roads in on both sides, and these streets could become a attractive area for outdoor living again, with kids playing ball games and cycling in the street, and all the sense of community that Clement Attlee had hoped for.

I cycled past a road sweeper van (renamed in *The Meaning of Liff*, the humorous dictionary, as a 'Vancouver'), bumped over a series of sleeping policemen (another fabulous description), and slowed to allow a blind man to cross the road at a zebra crossing (ditto). He continued carefully on his way, feeling along the pavement edge with his white cane, both of us concentrating on paying attention. Contrasting ways of exploring and making the invisible visible.

I love April. It took doing this project for me to realise that. Previously, I would have ranked it as a cold and blustery month, far down my list. But this year had changed my mind. The natural world bursts back into life in April, refilling me with optimism and recharging my batteries. Robert Browning pinned down the feeling long before me:

'Oh, to be in England
Now that April's there,

And whoever wakes in England
Sees, some morning, unaware,
That the lowest boughs and the brushwood sheaf
Round the elm-tree bole are in tiny leaf,
While the chaffinch sings on the orchard bough
In England – now.'

I walked around a graveyard accompanied by the songs of chaffinches, robins and blackbirds. My first red admiral butterfly of the year flitted past and a sprightly pied wagtail landed, wagged its tail up and down, then dashed across the grass in search of food. The trees were full of leaves, and the grass was a mosaic of dandelions and daisies. Those flowers are so common that I almost did not register them or note them as anything interesting. But had I never seen them before, it would have seemed spectacular. So I made the decision to appreciate them.

Dandelions go by many nicknames such as blowball, cankerwort, doon-head-clock, milk witch, Irish daisy, monks-head and priest's-crown. Their diuretic properties are well-known, hence one of its French names being *pissenlit*, or 'wet the bed'. The French also gave us our word dandelion, for the leaves resemble the teeth of a lion, or '*dents de lion*'. It has a more practical name in Chinese which translates to 'flower that grows in public spaces by the riverside'.

As for daisies, the name 'day's eye' comes from the way they open every morning. Chaucer called the flower the 'eye of the day'.

'Of all the floures in the mede,
Than love I most these floures white and rede,
Soch that men callen daisies in our town;
...
That well by reason men it call may
The daisie, or els the eye of the day,
The emprise, and floure of floures all.'

I moseyed around the graveyard for a while, discovering that one of the inventors of the steam locomotive (and a keen Cornish wrestler) rested in a pauper's grave here, while one of his colleagues had amassed enough money for a large family tomb a little farther along.

Among the many centuries' of graves were three Protestant martyrs, each burnt at the cross. Such religious fervour is hard to imagine in

Britain these days. We are more diverse and tolerant than ever before, despite the ongoing hysterical culture wars over our differences. I wondered what the angry folk back in the day would have made of some of the other churches I saw within this one square kilometre, including a Methodist chapel converted into a Vietnamese Buddhist Meditation Centre, The King of Glory Assembly, an old cinema that was now the Evangelical Alliance and Assemblies of God UK, and a corrugated iron warehouse for The Redeemed Christian Church of God.

I dropped down into the town centre, a hotchpotch of old buildings and alleys mixed with modern 'improvements' – shopping centres, roundabouts, and one-way systems. There was an 11th-century church by the river, plenty of rubbish, and, mysteriously, a neat trinity of tins of chickpeas, new potatoes and tomato sauce. You could relax on two benches, but someone had upended a third into the river. A tall elm tree, as featured in Browning's ode to April, cast shade over the benches, one of the rare survivors of the Dutch elm disease.

It was market day, and the high street was busy with trading stalls and pedestrians browsing racks of clothes, vaping supplies and bowls of vegetables. Bunting fluttered overhead, criss-crossing the street with red-and-white St George's cross flags. I liked seeing the display for our saint's day, despite England's general reluctance to celebrate itself.

This area had been a fashionable place to live until the 17th century, but seemed to have faded since then. There have been beer houses here,

water mills, claypipe makers and greengrocers, built over the ages along the Roman road that forded the river. Today's assortment of small businesses served residents from every corner of the Roman Empire and far beyond. There were shops selling produce from Transylvania, the Philippines, Turkey and more.

I have missed being in distant lands this year, but today gave me a bustling reminder of marketplaces around the world, listening to foreign tongues, and the timeless pleasure of people watching. Opposite an African grocery was a café whose menu board offered eggs florentine and mashed avocado, while next door's menu was resolutely old-school:

- Egg, bacon and sausage: £3.20
- Egg, bacon and mushrooms: £3.20
- Egg, bacon and bubble: £3.20
- Egg, bacon and beans: £3.20
- Egg, bacon and tin tomatoes: £3.20
- Egg, bacon and black pudding: £3.20
- Egg, bacon and onion rings: £3.90

The price hike for onion rings intrigued me…

Tables outside the church had been laid with polka-dot cloths, bunches of flowers and stands of cakes, tempting me to stop for coffee and scones in the sunshine. An elderly lady watched me go through the rigmarole of photographing my scone before scoffing it in far less time than the photo had taken me. I enjoy taking pictures, though, so it never feels like a hassle.

'Are you a photographer, or are you learning?' she asked.

'I think you're always a learner, aren't you?' I replied.

I pedalled on past the shopping centre and outlet stores towards the industrial estate behind it. I enjoy places like this, with all sorts of banging machines and weird equipment that I don't understand. I peered around inquiringly, on my way to the open land beyond the town's enormous, redundant gasometer. They are still a familiar sight, with one of the most famous gasometers overlooking the Oval cricket ground in London. But since the discovery of natural gas beneath the North Sea in 1965, they have become obsolete as gas is now pumped directly into homes.

The river flowed out of the town centre, past the gasometer, and over the ford where hermits once helped travellers cross. It morphed unnoticed under a modern bridge, from a small river into a broad creek, flowing slowly out of today's grid square towards the tidal marshes I explored some months ago.

I began the day on the hilltop estate and journeyed through the town centre's layers of history and language. Beyond the margins of welding businesses and repair shops lay an old mill converted into a 'luxury collection of apartments where aspirations are brought to life'. Segregating a map into kilometre squares is an arbitrary way to see a place. Within each square, everyone's lives were also divided and separated, often with little in common but a postcode. I can't reach for conclusions after such brief visits. But I found these snapshots of what life is like for different people in my district was helping me make more sense of my own life. As well as that, April is a month to look for hope in every corner and to turn your face to the sun whenever you need a reminder that summer always follows winter.

BLUEBELLS

'I do know how to pay attention, how to fall down
into the grass, how to kneel down in the grass,
how to be idle and blessed, how to stroll through the fields,
which is what I have been doing all day.
Tell me, what else should I have done?'

Mary Oliver

'Get out of the bloody field!'

'I'm on a bloody footpath!' I yelled back, both because I was angry and because the man leaning out of his 4x4 window was far away on the road.

It was an ineffective, hard to hear argument, so I just turned my back on the irate driver and continued following the path across a grassy field. I hate any form of confrontation – even a cross tweet upsets me all day. But this one particularly annoyed me because I was on a public footpath.

I would have understood the landowner's anger had there been no right of way and I was trampling crops, tearing up the land on a motorbike, dropping litter or worrying livestock. But his assumption that he had more right than me to the earth, wind, sun or sky irritated me.

We all need to access the natural world for our enjoyment and health, and if enough of us develop a connection with nature we might be able to reverse its destruction. But our history and laws have put so much of the countryside in the hands of so few people, that we have allowed a culture to establish where going for a walk is seen as invasive or damaging.

This 'get off my land' ticking-off put me in a blue mood when I should have been enjoying the clear blue skies and the bloom of bluebells. And it was a shame, because I could see that a lot of trees had been planted here – something that always lifts my spirits – so the landowner and I probably had far more in common than the gulf of our shouting match suggested. Had we chatted congenially and disagreed agreeably, the two of us would more than likely have ended up cheering for trees but feeling frustrated at government feet-dragging.

For example, one tree-loving landowner told me they tried to plant 200 acres of woodland, aided by receiving a grant that didn't make the venture profitable or even balance the books, but at least made it manageable for them to do the right thing for the land. But they were then told to apply for planning permission to plant the wood. By the time it came through, policies had changed and the planting grant had been withdrawn, leaving them with tens of thousands of tree whips sitting in their greenhouses. It is so frustrating to hear stories like this.

However, I rode out on this fresh spring morning to look for bluebells, not to get in a grump. And so into the woods I went. I was not disappointed. Bluebells are a highlight of my year. Their season is one of the most delightful times in England, with pockets of shaded woodland overflowing with carpets of flowers. Bluebells grow slowly and take a long time to establish, so woodland where they flower is usually ancient.

Because they flower early in the year, bluebells provide important nectar to pollinating butterflies, bees and hoverflies. I was taken aback by their abundance and vibrancy, such extravagant beauty in our generally muted landscape. Yet Britain has half of the world's bluebells. They are one of our natural crown jewels.

A bluebell's many colloquial names include cuckoo's boots, granfer griggles, witches' thimbles, lady's nightcap, fairy flower, and cra'tae – crow's toes. Fables about bluebells abound. Bluebell woods were said to be enchanted by wicked fairies who lured and trapped people. Should you ever hear a bluebell ring, you would be visited by a menacing fairy and would soon be dead. Picking a bluebell is also hazardous, for fairies may then lead you astray, leaving you to wander lost for all time.

Less sinister notions held that bluebells were symbols of humility, gratitude and love. Turn one inside out without damaging it, and you would win the person you loved. Wear a wreath of bluebells and you could speak only the truth.

Bluebells had many practical uses in olden times, with their sticky sap used for gluing books and sticking the fletching feathers onto arrows. Elizabethans crushed the bulbs to get starch for the ruffs of their collars and sleeves.

Every grid square in Britain has, over millennia, been changed and managed. Nothing is wild, nothing is untouched. But the plus side is that everywhere comes with layers of stories like these woven into it. Somebody has walked each square before me, people whose tales I would be intrigued to know.

With so much history in the landscape, I perpetually risked opening one of those Russian dolls that go on for ever (or up to fifty-one, actually, which is the world record for Matryoshka dolls). For in every grid square walk the ghosts of Romans and Victorians. The hidden histories of medieval farming practices and Neolithic burial beliefs lie in every field, never mind the aeons of evolution in every plant, and the billions of years of geology beneath every footstep.

And this is before I even get onto all the tangential ideas that each square brought to my mind; memories of other places, plans for future journeys, books I have read, random trivia. In other words, my single map was in danger of unfurling into a lifetime of work and wonder.

Louis MacNeice put this feeling better than me in his poem 'Snow'.

'World is crazier and more of it than we think,
Incorrigibly plural. I peel and portion
A tangerine and spit the pips and feel
The drunkenness of things being various.'

Leaving the bluebell wood, I stepped over a ditch full of delicate mauve cuckoo flowers and yellow cowslips, then followed a footpath across grassy fields in the sunshine. To my delight, I saw my first swallow, back on my map after a gigantic migration from Southern Africa.

They had arrived! I cherish the swallow's chirping song and acrobatic flying that 'catch at my heart and trail it after them like streamers', as Annie Dillard wrote in *Pilgrim at Tinker Creek*, observing the birds around her home in Virginia's Blue Ridge Mountains. A swallow has even been spotted in Antarctica, making it the only one of the world's 6,000 species of songbird to have reached all seven continents.

Because they only eat flying insects, the food supply for swallows dwindles at the end of the summer. So, after a too-brief visit, they head south once again, nonchalantly crossing the Sahara desert at a rate of around 200 miles a day. I doff my cap at this, remembering my own laboured efforts in deserts, whether cycling in the Nubian or Taklamakan, running marathons in the Sahara or heaving a cumbersome cart through the Empty Quarter in Arabia. Next year, the swallows will return to us again, with almost half the pairs returning to the very same nest. Seeing that one swallow today had made my spring.

Bird migration baffled our ancestors, and even Gilbert White struggled to believe in it. Undertaking a 12,000-mile round trip in search of food sounds outlandish, so alternative theories prevailed for the swallow's mysterious appearance then disappearance. Humans wrestled with the notion that small birds could accomplish navigational journeys that were beyond our heroic explorers. Perhaps the birds hibernated, people guessed, burrowed into the mud at the bottom of ponds, or even flew to the moon and back?

It was peak lambing season now and young lambs pranced in the pastures and butted their mothers for milk. The ewes eat grass and the suckling lambs enjoy a mixed diet of grass and milk that helps them put on 300 grams a day. The popularity of eating sheep meat (mostly lamb) is declining in Britain, though it is still higher than in many other countries, perhaps being culturally engrained by the historical importance of our wool trade.

Wool exports drove the British economy from the 13th to the 15th century. Wool was known as 'the jewel in the realm'. All Englishmen had to wear a wool cap to church on Sundays to support the wool industry. Even today, the Lord High Chancellor in the House of Lords sits on a woolsack as a reminder of the economic importance of the wool trade. It was the desire for wool that led to the notorious Highland Clearances in the 18th and 19th centuries, which brought eviction and famine to much of rural Scotland. Populations were forcibly removed from the land to create more space for sheep. The clearances destroyed the clan society and led to rural depopulation and mass emigration.

Dominating today's grid square beyond the fields of sheep was a transmitter station, a colossal metal spike thrusting high into the sky and dotted with dishes to transmit radio signals across the land. It was a sensible location, for I could see right across my map and far beyond. The green landscape fell away in every direction, dissolving into the hazy horizon. I always considered my map to be suburban, but from up

here it was evident how much countryside stretched across the region.

I crossed a busy motorway via a footbridge, stared down at the hypnotic rushing traffic for a while, and then went into some woods to search for what I was pretty sure would be the highest trig point on my map.

I like wild land squashed up against motorways, secret hideaways unnoticed by the thousands of people thundering by just metres away, faster than cheetahs, listening to Radio2CapitalRadio4ClassicFM and drumming on their steering wheels. The subtle shades of the woodland, its stillness and birdsong, all lulled me after the roar of metal, the rush of primary colours and the tang of burning fuel.

The wood was mostly oak and beech with a carpet of dog mercury beneath, but there was also a stand of Scots pine trees, with a bare floor of needles, that reminded me of Castile, minus the chirrup of cicadas. I once spent a month hiking through Spain with no money and only my scant violin busking skills with which to earn my next loaf of bread. I often sought siesta shade beneath trees like these. It was one of the best times of my life.

I found my trig point on the far edge of the wood, enveloped in brambles beside a flinty ploughed field. It took three attempts with my camera's ten-second timer to successfully capture my dash, climb and celebratory pose, but I got there in the end, arms aloft and grinning on top of the trig point. The high point of the map – my Everest for the year – and a grid square of sunshine, swallows and bluebells. Today was definitely worth smiling about, and no irate landowner begrudging me enjoying our ancient rights of way was going to spoil that.

SUBURBS

'Nothing, like something, happens anywhere'

Philip Larkin

Much of today's square was taken up by stuff that loosely lumps together under the heading of 'infrastructure'. Railways, roads, roundabouts and railings. Big metal things. Corrugated sheds. Padlocks. Pylons. Pick-ups with orange hazard lights. Men in hard hats. Things I don't understand but that I know are important. All the 'Keep Out' signs on this grid square were definitely for the best.

I tried to get a closer look at a 400kV electricity substation, but its mysteries were obscured by rings of trees because, between 1968 and 1973, an admirable 725,000 tall trees, 915,400 smaller trees and 17,600 ground cover plants were planted to screen substations across the land.

My limited interest in infrastructure exhausted, I followed a cycle path alongside the dual carriageway, dodging broken bottles amid the traffic roar. The smells of warm tarmac and diesel brought back fond memories of cycling the world's highways. I peered down from a bridge at an overgrown pond, thick with slime and dotted with traffic cones. Then I turned off at a slip road and rode into a town. There were large, detached houses at the top of the hill, and the homes became smaller and closer together as I freewheeled down towards the town centre. A pony and trap cantered by, ridden by two young lads in vests, and trailing a patient line of backed-up traffic in its wake. I left the main road to go and cycle around some residential estates.

Over the course of this year, I'd always enjoyed visiting grid squares that most approximated wild countryside. And I also liked the busy towns brimming with human life, being equally intrigued by mansions and poorer areas. Today I was bang in the middle, riding through street after street of suburban homes.

The uniform brick houses stood in perfect order, all displaying neat and tidy front gardens. I didn't see anybody outside or walking along the streets. Only occasional cars drove past, slowed by regular speed bumps. Each residence had practical driveway parking, accommodating cars that were washed and polished with regularity.

This was an area buffed with aspirational pride, embodying a striving for 'familiarity and endurance, security and safety', as Karim said in *The Buddha of Suburbia*. It was unusual to explore a built-up area of my map without seeing anybody else. Everything felt eerily quiet as I took a photo of a garden ablaze with tulips next to a neighbour's covered in paving and power-hosed spick and span.

Those tulips had come a long way. Originally growing wild in the valleys of the Tian Shan mountains in Central Asia, tulips have been

cultivated for a thousand years. They became popular in the West in the 16th century after diplomats saw them in the Ottoman court in Constantinople. Their popularity continued to grow until tulips eventually became a valuable, tradable commodity.

During a mad period in the Dutch Republic known as Tulip mania, the craze for the flower boomed and prices soared. The madness peaked with a single *Semper Augustus* bulb whose value was 'sufficient to purchase one of the grandest homes on the most fashionable canal in Amsterdam, complete with a coach house and a 25-metre garden'. But the market collapsed spectacularly in February 1637, and after the crash the tulip returned to being just a nice flower. Tulip mania was the first recorded asset bubble, leading on to the more recent dotcom bubble and the US housing bubble.

These suburban streets were comfortable, safe rewards for working hard, living sensibly and settling down. All good things, of course. But I wanted to find less order, more nature and more surprises, so I made my way towards the fields behind the cul-de-sacs. A clear boundary marked the end of the development and the beginning of agriculture, or the other way round, depending on how you looked at it.

Once I got out into the fields, I saw a similarly sharp boundary between the farmland and the remaining fragments of forest on the edge of my square, a wood that had become a farm and then a suburb. The fields were flat, bare earth, with pylons down the middle and dog walkers around the fringes. They were hard fields to love, so I was pleased to get into the trees.

The woods made me feel better, as they always do. I came across the framework for a treehouse in an oak tree, the ground beneath it strewn with Strongbow Dark Fruit Cider cans, one of the preferred drinks for local litterers. Over the ages, there have been various fashionable periods for treehouses. The Roman emperor Gaius Caesar, nicknamed Caligula, had one in the branches of a plane tree in the Alban hills. The fabulously wealthy Medici family rekindled the trend in Renaissance Italy, and they became popular again in England's formal gardens in the 17th century. Whoever had been working on this treehouse was having fun building it in the 21st century, the latest golden age for treehouses. I made one a few years ago and enjoy climbing up there on summer days when it's too hot to be in my shed.

I headed deeper into the wood, looking for a comfy spot to brew a cup of coffee, and stumbled upon a homemade BMX track, its berms and bumps constructed from mud and branches. A discarded dandelion and burdock bottle was a flashback to my childhood (£1 for two litres containing 20 teaspoons of sugar). The empty lager cans, torn and burnt pages of school books, and shredded air-gun targets suggested all sorts of youthful antics beyond riding bikes.

The BMX bandits were all at school this morning, so I had the clearing to myself. I sat in the sunshine and boiled water on my small gas stove. I enjoyed this ritual not so much for the coffee, but for the *caesura*, the enforced pause and stillness of waiting for the pot to boil. My weekly forays often felt hectic. I was constantly alert, taking photos, making notes and scanning the landscape. I don't seem to be able to do anything without it descending into a hurry and a challenge. Sitting still with a cup of coffee was my antidote to that. And if you sit still in a wood, you will almost always be rewarded.

I heard the magnificent drumroll of a great spotted woodpecker somewhere beyond the white carpet of star-shaped wood anemones. When there is a covering of these small flowers, you know you're in ancient woodland, even if little of it remains. Wood anemones rarely set seed, and spread only six feet per century through their roots, so this carpet of flowers had been a long time in the making.

Jake was by himself, too.

The woodpecker smashing its head against a tree was in search of lunch or love. Woodpeckers don't claim territory to impress mates by singing, as many birds do. Instead, they make themselves known by drumming on dead trees with their beaks. It sounds painful, but their skulls are cushioned with tiny pockets of air and supported with strengthened bone tissue. This shock absorption allows them to drum on wood, peck for insects and excavate nest holes.

I finished my coffee and rode home in the warm sunshine. It was the sort of day where you leave home in a jacket and gloves, then return in a T-shirt. The sort of grid square that begins boring but ended up taking me somewhere I'd never been and getting me out to enjoy some miles on my bike.

I need regular doses of exercise and fresh air to help me to unwind and get through each week. My life sometimes seems to consist of filling the dishwasher then emptying it again. I envy those neighbours in the suburbs who can enjoy the familiarity and routine of regular life. It tends to drive me a bit nuts. These weekly outings were dampening the madness that daily life incites in me, and helping me to accept and appreciate the gentle curiosities of my local map.

CUCKOOS

'Each town should have a park, or rather a primitive forest, of five hundred or a thousand acres, where a stick should never be cut for fuel, a common possession forever, for instruction and recreation.'

Henry David Thoreau

I was back on the marshes where I'd begun my journey almost half a year ago. I liked it out here. A town lay in the distance, its prominent wind turbines turning steadily. I preferred these empty corners of my map, the ignored and forgotten places. I was drawn to their anonymity and the distance they put between me and all the things I 'should' be doing in life, the sort of things I imagined everyone else seemed to tolerate or enjoy but that left me frustrated and wanting to be elsewhere. Perhaps that explained my recent indifference to the orderly suburbs. I felt I was the only one who thought this way, yet I'm sure there are other people on my map also quietly pounding the walls and howling at the moon. But I didn't know any, and I never saw anyone else in these empty places.

After the recent sunshine, blustery weather had returned, and I donned thick gloves that I thought I'd put away for the year. I pedalled out through rush hour, slipping past cars hooting their frustration in traffic jams, including a gleaming baby-blue Bentley. Cities with enlightened transport planning systems, such as Hong Kong, Zurich or Stockholm, try hard not only to expand public transport access but also to end the blight of single-occupancy vehicles. A hallowed motto of urban mobility planning is that 'a developed country is not a place where the poor have cars. It's where the rich use public transport.'

I felt myself decompressing as I rode beyond the traffic into the countryside. The crops were thickening and greening, and I was delighted to hear my first cuckoo call of the year. Cuckoos were once part of the soundtrack to an English summer, but I rarely hear them these days as three-quarters of them have been lost since the 1980s. Cuckoos' habitats have to match the meadow pipits, reed warblers or dunnocks that unwittingly raise the young cuckoos in their nests, so their fate is intertwined with their hosts'. Agricultural practices are contributing to the decline of all these species, as birds are pushed out of the farmed countryside into what remains of wilder heath areas and uplands.

Cuckoos are brood parasitic birds who save themselves the bother of rearing offspring. By sneakily laying eggs in other nests, thus outsourcing all future childcare, they can allocate extra resources to producing more eggs. The host's offspring, meanwhile, have to compete with the imposter baby cuckoo, which even goes as far as trying to evict the host's eggs or kill their hatchlings. The host mother may be only a tenth the size of the young cuckoo, but the gargantuan cuckoo chick remorselessly demands food from her. Nature seems cruel, but she balances her books far better than we do.

I passed through a kissing gate and admired an old timber barn set on mushroom-shaped staddle stones to raise stores off the ground and protect them from mice and damp. Disappointingly, kissing gates don't derive their name from romance but because the hinged gate 'kisses' both sides of the enclosure. This hasn't stopped the tradition of requesting a kiss from the person who follows you through the gate. As usual, however, I was by myself, so I just hoisted my bike overhead (in a muscular way that would no doubt have driven any companion to kiss me) and then passed through the gate to the marsh on my own.

Marshlands are one of our most endangered landscapes. They are at risk from rising sea levels and could begin to disappear within the next twenty years. As well as global warming, they are also threatened

because they're low-lying and accessible, and therefore tantalising for 'development'. Besides their own intrinsic value, salt marshes protect against erosion and flooding, and their complex ecosystems are important for nature.

I followed a chalky, bumpy track across the marsh between two reed-lined drainage ditches. A reed warbler, recently arrived from Africa, burbled its lively song. I couldn't spot it in the undergrowth, but my trusty Merlin app identified its call. Reed warblers weave immaculate basket-like nests that are suspended and concealed in the reeds. Even so, they are not safe from cuckoos, and 10 percent of reed warbler nests each spring are home to an imposter. Fortunately, because warblers raise multiple clutches of eggs each year, cuckoos are not too damaging to their overall populations.

A heron descended like a paraglider, drifting for about 200 metres without flapping its wings, then braking hard and landing beside a flooded ditch to survey the landscape with its haughty, atavistic gaze. I stood still and watched it while a ship on the distant estuary appeared to drift past a field of cows. A lark sang overhead and I eased in to the sedate pace of marsh life.

I passed a fence post covered in bird droppings, a useful lookout spot for a marsh harrier on this low, open terrain. They too nest in these reedbeds, and hunt for frogs, mammals and small birds. Their courtship involves remarkable aerial displays, with couples locking talons in the air as they tumble from great heights. Marsh harriers were once very endangered, but their numbers are increasing again thanks to habitat restoration, the banning of certain pesticides, and a clampdown on illegal shooting and egg collection.

The footpath continued through a docile herd of silky Hereford cows. They are the product of systematic breeding practices begun by Benjamin Tomkin in Herefordshire, way back in 1742. All 5 million pedigree Herefords today are descendants of a single bull, the son of a cow named Silver, and two cows called Pidgeon and Mottle. They are now one of the most widespread breeds in the world.

'Good morning, ladies,' I called out, though they mostly ignored me and kept munching the grass. 'Don't be moody. Would you like to be in my book? The steaks are high, but I'll milk it for all I'm worth. Are you not amoosed?'

And so on.

The disused canal beside which I had sat on my first outing sliced across the marsh. Its construction began in 1800 with great fanfare, but the era of the railway arrived before the canal was completed, rendering it instantly obsolete. The canal opened twenty-five years after work began, coming in at eleven times over budget and never being particularly commercially useful. Today, the towpath is a peaceful conduit for joggers and cyclists, and its overgrown waters are home to populations

of moorhens, frogs, Lucozade bottles and deflated birthday balloons.

I zipped cheerfully along the towpath, now part of Sustrans' National Cycle Network that was established to encourage us all to ride and walk more. Launched with a £42.5 million National Lottery grant, the NCN is now used for more than 786 million cycling and walking trips per year, on 12,739 miles of signed routes, of which 5,220 miles are traffic-free paths like this one.

A pair of feisty reed warblers almost crashed into my bike while I was riding at speed. They were busy squabbling or courting on the wing and paid scant attention to the giant human on his beloved blue bicycle. In a vivid split second, I watched a bird fan open its wings, slam on the brakes *Top Gun* style, and veer away from my spinning spokes just in time.

It was glorious out here today. Spring was in full cry, a season that comes along slowly, slowly, then all at once. I was in a country that cared, to a greater degree than most nations, about encouraging cycling and walking. I had heard my first cuckoo of the year. And I had chatted with some cows. What more could I ask?

THE
LIGHT

HALF

BELTANE

MAY

CLOUDS

'At any rate, spring is here, even in London N.1, and they can't stop you enjoying it. This is a satisfying reflection. How many a time have I stood watching the toads mating, or a pair of hares having a boxing match in the young corn, and thought of all the important persons who would stop me enjoying this if they could. But luckily they can't… The atom bombs are piling up in the factories, the police are prowling through the cities, the lies are streaming from the loudspeakers, but the earth is still going round the sun, and neither the dictators nor the bureaucrats, deeply as they disapprove of the process, are able to prevent it.'

George Orwell

First stop: the skate park. It was a school morning, so I had the place to myself. I amused myself for a while riding up and down the ramps on my bike, except for the steepest one, which was too scary. The fields were still damp with morning dew when I'd finished mucking about, sparkling with a million tiny diamonds.

The fleeting beauty of these shining droplets, each with an upside-down image of the world hanging within it, led to early superstitions. It was said that the dew on the first of May had magical properties that bestowed a flawless complexion on anyone who washed their face with it. A rhyme suggested that a maid who rises early on May Day 'and washes in dew from the hawthorn tree, will ever after handsome be'.

The diarist Samuel Pepys recorded his wife travelling out from London to the countryside in Woolwich at the end of April 1667, ready to collect May dew, 'which Mrs Turner hath taught her is the only thing in the world to wash her face with'.

Variants of the formula suggested the May dew had to come from ivy, or perhaps the grass beneath oak trees. But everyone agreed it worked best at sunrise, so some people stayed out all night to catch the first dew. There may have been additional motives at play here, with the Puritan Philip Stubbes harrumphing in 1583 that out of all the young women who spent May Day eve in the woods, 'scarcely the third part of them returned home again undefiled'.

The roadside verges were growing thicker and greener every week. Oxeye daisies waved in the draught of passing cars. I nibbled a few leaves of garlic mustard, also known as Jack-by-the-hedge, and a nice addition to salads or when cooking fish or meat. In Somerset folklore, it was said that rubbing the fresh leaves onto your feet relieved cramp, though whether or not they were then added to the salad is unknown.

Another abundant plant this week was *Galium aparine,* better known by kids as cleavers, clivers, catchweed, stickyweed, sticky bob, sticky bud, sticky willy, sticky willow, stickyjack, stickeljack, grip grass, goosegrass, robin-run-the-hedge, bobby buttons or whippysticks.

Ancient Greek shepherds used the hairy stems of this plant to make rough sieves to strain milk. The botanist Carl Linnaeus reported the same thing in Sweden, while elsewhere in Europe the dried foliage was used for stuffing mattresses.

I still haven't grown out of finding it amusing to drape strands of stickyweed onto friends' backs when out walking. A slap on the back and a hearty 'mate, it's lovely to be out here with you' is all it takes to do the deed and transfer the weed.

Today's grid square was a land of expensive houses, double garages, and lawns with fountains and sculptures. One garden had a tall monkey puzzle tree. Next door was a splendid cedar. They always remind me of the flag of Lebanon and my time cycling through the mountains there. When Lebanon gained its independence from France in 1943, the new country adopted a white flag with two red stripes and a green cedar tree. The white represented peace and purity, while the red stripes were a reminder of so much bloodshed. This, though, was only the latest in a line of thousands of years of flags in the region.* It is useful for your sense of perspective to reflect on the impermanence of everything we consider fixed, such as enormous cedar trees, landscapes, national flags or, indeed, nations.

Meanwhile, the monkey puzzle tree next door to the cedar hailed from Chile. They were first brought to Britain in 1795 as an ornament. Their distinctive spiky branches have defended the tree from grazing animals for 200 million years, though these days it is jays and squirrels that feast on the nuts, rather than dinosaurs, and the monkey puzzle is now endangered in the wild. A hefty 8,400 miles away from Lebanon and its cedar-tree flag, Chile sports a white and red horizontal stripe and a blue square with a white star on its flag. Once again, it wasn't always thus, and I wonder what will come next.**

Flagging from all this vexillography, or flag design, I rode into a village with old brick cottages and grand houses clustered around two appealing pubs ('It's too early, Al. It's too early'). One house was called The Old Chemist and another The Old Butchers, reminders of the changes in village life over the centuries. Today's occupants were more likely to work in an office in the city, have meat delivered from a supermarket, and buy medicines online from a warehouse on a distant grid square.

An antique-watch restorer had a studio in the village, and he shared my blossoming interest in random stuff. A note in the display window

**Five millennia ago, the flag was blue and red. Then came the Tanukh tribe's white-blue-yellow-red-green vertical banded flag, followed by the Abbasid Caliphate's plain black one. At other times between then and now: a white flag with a golden cross; a banner of diagonal white and red with a laurel wreath; a plain yellow flag; a yellow flag with a left-facing crescent moon; a blue flag and a right-facing crescent; a red flag with a white moon and star; a white flag with a green cedar tree; a French Tricolour with a green cedar, another with a black one, and a third with a brown trunk and green leaves.*

***The first flag flown in what is now Chile was the Mapuche war flag, blue with a distinctive white star called a* guñelve. *That flag fell to the invading Spaniards' saw-toothed burgundy cross, and the conquerors ruled Chile beneath their national flag until 1812. Upon winning independence, Chile adopted a blue, white and yellow tricolour, nicknamed* Patria Vieja. *Over the next century, various modifications came and went until 1934, when they plumped for the design still used today.*

asked, 'Does anyone know what these pliers were used for? Let me know if you do. Thanks, Mark.'

A peculiar spiky object had a note that asked simply, 'WHAT IS THIS?'

The mystery had been solved, and an update on the note gave the details. 'Thanks to a customer, I can now tell you that this is a larding needle used to inject fat into lean meat. Look up larding and barding. Best wishes, Mark.'

OK, Mark, I shall.

Larding beef means to 'artificially marble the meat with fat. The fat is introduced in the beef's cut using a larding needle. The fat injected can be beef fat or more likely pork fat. Barding beef involves wrapping the meat in two or more strips of fat which have the purpose of protecting the meat from drying. There are people who confuse larding beef with barding beef but the difference is very clear.'

Very clear.

I pedalled out of the village and away from the madness of becoming engrossed in every single thing my eyes fell upon. A grassy byway led towards a triangle of woodland by the motorway. Mighty oak trees sported their first flourish of unfurling green leaves, and a little wren blasted out its surprisingly loud song. For their size, wrens are ten times louder than cockerels.

Birds can produce much greater sounds than we can because they use the syrinx at the bottom of their windpipe. It is surrounded by a resonating air sac chamber that amplifies the sounds, unlike our larynx. The syrinx is also often double-barrelled, which helps to produce more complex sounds than we can manage. It would be like us being able to play both a trumpet and a tin whistle at the same time, with each lung blowing a separate instrument.

A footpath led through the wood and crossed the motorway via a narrow pedestrian footbridge. The Samaritans charity had fixed signs to the bridge to try to reach anyone feeling suicidal. 'Talk to us, we'll listen. Whatever you're going through, you don't have to face it alone.'

In 1953, Chad Varah, a vicar and writer-cartoonist, began answering phone calls from suicidal people on a helpline he set up. He worked all day, then stayed up late into the night taking calls in between writing scripts and cartoons. The Samaritans has grown so much since then that it now responds to a call for help every ten seconds, which I find both inspiring and heartbreaking.

Over the motorway, heathland stretched towards the tall fences of a quarry. Once again, I realised that I was hurrying for no reason, so I chose to notice that the sky was a deep blue and that the white clouds were scudding along in high-altitude winds. I lay down in the grass and gazed up for a few minutes. The sun felt pleasant on my face

as I studied the clouds and imagined the shapes I could see, just like Hamlet in his famous exchange with Polonius.

'Do you see yonder cloud that's almost in shape of a camel?' asked Hamlet.

'By the mass, and 'tis like a camel, indeed.'

'Methinks it is like a weasel.'

'It is backed like a weasel.'

'Or like a whale?'

'Very like a whale.'*

The quarry was a chasm gouged twenty metres down into the earth, equivalent to a mind-boggling number of sandcastles, excavated for building sand, top soil, southern side sand, equestrian sand, reject sand and silica sand. To pick just a couple of these specialised categories of dirt, silica sand has been used in glass-making for thousands of years and was a vital component of the industrial revolution. Topsoil, the first layer to be excavated in a quarry, is high in organic matter and invaluable for growing plants. Indeed, it is one of the most important and underrated components of life on earth.

Almost 4 million hectares of soil are at risk of compaction in England and Wales, and 2 million hectares are at risk of erosion. Intensive agriculture has caused arable soils to lose up to 60 percent of their organic carbon, and soil degradation costs us over £1 billion every year. If we continue to neglect soil, losing it far faster than it is created, then the world could run out of topsoil within sixty years, with disastrous consequences. To repair the soil upon which we depend, we must shift our farming systems to ones that use less fertiliser, reduce ploughing, and use cover crops and crop rotation.

Looking down into the quarry, it jarred that a manmade hole had literally reshaped the landscape. I am familiar with towns and roads, so that sort of reshaping doesn't look shocking to me anymore. But this vast hole on my grid square prompted a small nudge in my awareness about our endless desire for development and the impact this has had on every square of my map.

**If you also enjoy clouds, like Polonius, Hamlet and myself, you can join the Cloud Appreciation Society (annual membership, £25).*

GREEN MAN

'I am alarmed when it happens that I have walked a mile into the woods bodily, without getting there in spirit. In my afternoon walk I would fain forget all my morning occupations and my obligations to Society. But it sometimes happens that I cannot easily shake off the village. The thought of some work will run in my head and I am not where my body is – I am out of my senses. In my walks, I would fain return to my senses. What business have I in the woods, if I am thinking of something out of the woods?'

Henry David Thoreau

My childhood bedroom overlooked a village green, and I have been fond of those open spaces ever since. My brother and I used to hang out there with our friends. It was our amphitheatre, the scene of day-long rugby matches, and a cricket pitch with the twin hazards of horrific bounce after cows had been herded across the wicket, and the risk of a lost ball if an exuberant shot sent it flying into the garden of the grumpy man who lived in the cottage in the centre of the green.

Given that it was early May, it was apt that the pub on today's charming village green was called The Green Man. Appearing in various guises over time – usually a green head sprouting leaves and foliage – the Green Man used to be a central figure in May Day celebrations.

His origins are murky, but he has been carved in churches and buildings for a thousand years as a symbol of spring's rebirth. The Romans had similar figures, as seen, for example, in Nero's Golden House palace. Bacchus, god of wine, nature and harvest, was often portrayed as a leaf-crowned lord, so he might be the ancestor of our Green Man.

The Gaelic festival of Beltane was a forerunner of today's numerous worldwide celebrations of May Day. It was a community celebration of summer's return. The origin of the word Beltane is 'bright fire' and, as always, bonfires played an important role in the rituals. Revellers danced around purifying flames to welcome the lighter half of the year after the long winter. When farmers led their animals out to spring pastures, they made sure to drive them between two fires to bring good luck.

I was cheered not only by the bucolic village green, but by the fact that the pub would be open in a couple of hours. So I cycled out of the village past an old red phone box that had been converted into a community library and larder ('Bring what you can, take what you need'), and past the village hall polling station for today's local elections.

I turned down a rutted farm track wedged between hedges that gleamed like emeralds. They were unusually thick and well maintained, and therefore brilliant for wildlife. We have half a million miles of hedgerows, only a fraction of what there was a century ago, but still an important ecological building block. The sides of the lane were flush with bluebells, a clue that there had once been forest here. Some

woodland species linger like lonely ghosts even once the trees have been felled, their roots grubbed out, and the forest turned over to cattle.

I blooming adore May. Everything was blooming and beginning again. A snippet of Larkin popped into my head.

'The trees are coming into leaf,
Like something almost being said.
Last year is dead, they seem to say,
Begin afresh, afresh, afresh.'

Little of what I learnt at school lingered long in my brain. But I once had an English teacher who made us memorise quotations and snippets of poetry, and I'm often grateful for it.

I shoved my way down an overgrown byway and emerged on a footpath through a wood that was spectacularly carpeted with bluebells in every direction. A beautiful place to pause and brew coffee. The path was pocked with hoof prints and I wondered why horses wear shoes, but other farm animals do not. Apparently, wild horses amble long distances over rough grassland, which toughens their hooves, similar to us going barefoot for the summer. But domestic horses don't exercise as much, and when they do, it tends to be on softer ground, so they need horseshoes to stop their hooves wearing down or splitting.

I pootled across a farm, down grassy footpaths, unlikely to accidentally veer off course with the abundance of passive aggressive 'Please keep to the path' signs plastered everywhere, reinforced by bonus 'No through route' signs. I was permitted to tread only the most slender of threads. But this was a morning for expansive enjoyment, so I chose not to feel unwelcome.

The air was warm but fresh after last night's heavy rain. It had been tipping down all week, culminating at last in one of those clear-the-air rainstorms when the sky darkens, bursts, rinses, and then the sun comes out to polish the new world. Hornbeam trees – the hardest wood in Europe, used by the Romans for building chariots – sported their new leaves above banks of greater stitchwort. As with most hedgerow plants, most of us wouldn't look twice at the pretty, star-shaped, white flower or know its name (me included, until Seek taught me). And yet, over the centuries, the small flower has accumulated a host of local nicknames, including wedding cakes, star-of-Bethlehem, milk maids, little dicky shirt fronts and daddy's-shirt-buttons, as well as snapdragon because of its brittle stems, and poppers or nanny crackers for the seed pods that explode on summer days when they are ripe. I wish I could travel back in time to when everyone knew the names, uses and stories of the plants in their community.

I noseyed around the grid square for a couple more hours before succumbing to the magnetic lure of the pub. I was exactly halfway through my year now: twenty-six weeks and twenty-six grid squares. As I sipped a pint of bitter on the village green, I reflected on building the habit of getting out into nature to see what changed every week. Documenting my experiences had helped immerse me in the turning of the seasons more than ever before. Once you take an interest in nature's calendar there is always something to spark your curiosity when you go for a walk.

The Japanese calendar was traditionally divided into twenty-four periods, starting with *Risshun*, 'the beginning of spring', and finishing with *Daikan*, 'the greater cold'. The year was also sub-divided into seventy-two micro-seasons, each lasting for roughly five days and marking the near-continuous changes of nature. They have melodic names such as *Kōō kenkan su* or 'bush warblers begin singing in the mountains'. Others that I like include 'Mist starts to linger', 'Frogs start singing', 'Dew glistens white on grass', and 'Bears start hibernating in their dens'.

I toyed with creating my own version of that calendar, according to what happens where I live over the span of a year. I could begin with today's delightful micro-season: 'Beer in the sunshine, summer on the horizon.'

BUTTERCUPS

'The world will never starve for want of wonders;
but only for want of wonder.'

G. K. Chesterton

You should sit in nature for twenty minutes every day, they say, unless you're too busy; then you should sit for an hour. I sat for a while on the bench on a small, triangular village green because I thought I was too busy to be doing this today. It was a cold and blustery morning, so I was wearing a hat and gloves again and hunkering down into my collar. I'd hung all the washing on the line before heading out, but now it looked like it was going to pour with rain. I was also in a bit of a grump because this square looked dull on the map. But a few minutes of stillness helped to settle me into a calmer mood and slowed my impatient mind.

A sign on the green said the village was supporting No Mow May, which explained why the grass was peppered with wildflowers. In Britain we revere short, stripy lawns. But the charity Plantlife urges us to enjoy the beauty and the wildlife benefits that come from allowing lawns, greens and verges to run a little wild for a month. After No Mow May, up to 200 species can be found flowering on lawns, including such rarities as meadow saxifrage, knotted clover and eyebright, as well as an abundance of daisies, white clover and selfheal. The longer you leave a lawn unmown, the wider the range of plants, while cutting the grass every four weeks generates the greatest quantity of wildflowers and nectar.

Cycling out of the village, my first stop was a field covered in buttercups. The flower's sap is toxic for humans and animals. Medieval beggars rubbed the irritant sap onto their skin to create sores, hoping to receive more sympathy and money. Over in Sardinia, they believed that anyone foolish enough to eat the poisonous flower developed a twisted grin before keeling over.

I remember holding buttercups under friends' chins at school to 'test' if they liked butter. It is a sweet game, based upon a unique layer of reflective cells in their petals. But the flower's Latin name, *Ranunculus*, nods to darker myths. Ranunculus was a boy who dressed in silks of green and gold. He ran through the forests all day, singing loudly to himself until the wood nymphs got fed up with him, turned the lad into a buttercup, and banished him to live in the open meadows.

In parts of America, the buttercup is known as coyote's eye. Legend has it that the coyote was busy tossing his eyes in the air and catching them. Details are hazy as to why he was doing such a thing, and how he caught his eyes if he couldn't see them. Suddenly an eagle swooped down and snatched his eyes. The blind coyote was forced to fashion fresh eyes out of buttercups.

I was photographing the field full of buttercups when a lawn mower (the job description, not the machine) stopped his van, wound down the window and asked, 'What lens are you using?'

It was an odd question, and not the sort you usually expect when a white-van man skids to a stop beside you and winds down his window. But I'm always eager to talk about lenses.

'A 24-70, f4,' I said. 'Not wide enough to be wide, not zoomed enough for the long shots. Not brilliant at anything, but quite a useful all-rounder. Maybe a bit like me.'

It turned out, however, that the man did not want to know about my camera equipment: he wanted to tell me about his own. He spent the next ten minutes pulling lenses from a camera bag on the passenger seat and describing them all. I liked his enthusiasm. He also explained why he enjoyed mowing people's lawns.

'I go to some pretty nice spots, so I always take my gear with me. There's loads of stuff to photograph and put online.'

We were approaching similar things in different ways. We both liked seeing new places, then photographing and sharing them. And we'd found jobs that allowed us to get outdoors and do what we enjoyed.

'What do you think about No Mow May?' I asked.

'Bloody stupid idea!' he laughed.

I wished him well and made my way into the woods to make coffee. Many of the ash trees had been felled to manage ash dieback and

slow the spread of the disease. Sprayed orange marks indicated those that were next for the chainsaw. It was sad to see so many condemned trees. Ash makes up 12 percent of our woodlands, so dieback is going to radically change our countryside for decades to come. It will cost billions and put strain on many species that rely on ash trees, including dormice, bullfinches, privet hawkmoths and nuthatches.

Finding a chain-sawed ash tree in a pleasant clearing of bluebells, I made the most of the tragic situation by manoeuvring a smooth white slab of wood to make a temporary coffee table. I even used the spirit level app on my phone to ensure it was perfectly level. Melchisédech Thévenot invented the spirit level in the 17th century. He was a librarian for Louis XIV and an amateur scientist. He is better known for his popular 1696 book, *The Art of Swimming*, one of the first books on the subject. It popularised breaststroke, and even Benjamin Franklin was reputed to be a fan.

As I fired up my little gas stove to boil water, I admired my handiwork and uncharacteristic attention to detail. Soon my moka pot was roaring and I turned off the flame to enjoy the sudden silence and the dark, strong coffee. Taking time to sip a decent cup of coffee out in nature had been a treat this year.

A freewheeling, no-handed downhill ride is another of life's delights (also on the list is picking blackberries, fresh snow, peeing off cliffs and jumping into rivers). I flew down through a tunnel of trees towards a village spread across the foot of the valley. The hills were unusually steep in today's grid square and one climb even got me out of the saddle and panting in granny gear.

I stopped to look at a field that was half wheat, half wildflowers. The environmental impact of feeding us all was apparent in the difference between the green monoculture and the haphazard variety of dozens of plants, including purple ground ivy and crosswort.

Meanwhile, at my feet, a loveliness of ladybirds (a fine collective noun) were going about lunch in their own way, foraging on a patch of nettles. There is a greater mass of insects on earth than humans, and yet they tread far more gently than we do (barring the occasional biblical locust swarm). Every day, humans burn ninety million barrels of oil. Every minute, we subsidise the fossil fuel industry by £8.8 million. And every second, we cut down a football field's worth of forest. We're being immensely stupid with the planet. Ladybirds, by contrast, nibble a few aphids and look cool.

The name ladybird is a contraction of 'Our Lady's bird', for early Christian art often depicted Mary wearing a red cloak, and the insect's seven spots symbolised her seven sorrows. The German word for a ladybird, *Marienkäfer*, translates as 'Mary's beetle'. In Hebrew, it is known as 'Moses' little cow', and its names in Irish, Polish and Russian all translate to 'God's little cow'.

I knew the nursery rhyme 'Ladybird, ladybird', but hadn't remembered the grisly lyrics.

'Ladybird, ladybird fly away home,
Your house is on fire and your children are gone,
All except one, and her name is Ann,
And she hid under the baking pan.'

A shorter, grimmer version of the rhyme concludes,

'Your house is on fire,
Your children shall burn.'

Goodnight children, sweet dreams.

I rode along a quiet lane and passed several people walking pairs of greyhounds on leads. They had come out from the racing kennels nearby to stretch their legs. Greyhound racing evolved from the 'sport' of coursing, or hunting hares, which is still practised but now illegal. I used to enjoy an evening at the dogs in Wimbledon before the stadium closed down, marvelling at the speed and intensity of the races. Greyhounds can run at 45mph – Usain Bolt hit just 27mph – and they are in the air for 75 percent of the time when in full flight. The graceful creatures were depicted in the tombs of Egyptian pharaohs and are the only breed of dog name-checked in the King James Bible.

I headed up to the top of the valley, restricted to narrow strips of footpath cordoned off from broad horse pastures by electric fences. The horses enjoyed more freedom on this map than I did. This was a beautiful spot, green and wooded in all directions. I could hear children playing at the school down in the valley, lorries rumbling along the distant motorway, and birds singing in the woods. The world was lush and full of life. I, too, felt full of life, restored, and in a much better mood than when I arrived here this morning.

I always liked being able to see a whole grid square in one expanse, appreciating the lie of the land and realising quite how large a square kilometre is when you travel slowly around it. You cover a kilometre in a mere thirty seconds on the motorway, and you'd have to run across five such squares to complete a single Parkrun, something that more than 350,000 people manage each week. So a kilometre is a relatively small distance, yet it is also as large as you want it to be when you slow down and immerse yourself in it.

SWIFTS

'Also, as I stand listening for the wren and sweltering in my great-coat, I hear the woods filled with the hum of insects – as if my hearing were affected – and thus the summer choir begins. The silent spaces have begun to be filled.'

Henry David Thoreau

I found an elevated spot where I could peep through the fence and look down on the new town being built across this blank grid square. Yet my map has never been blank. Even our brief history here stretches back hundreds of thousands of years to the prehistoric hand axes discovered nearby, tools once used to butcher animals and make clothes. I've heard that sort of fact so often that it didn't particularly astonish me. But learning that the axes were made by *Homo heidelbergensis*, an extinct species of archaic human, rather than by *Homo sapiens*, reminded me how rare it is for there to be just a single species within a genus (known as a monotypic genus). This is a dubious, lonely honour that we share with the dugong, narwhal, platypus, and not much else.

There used to be nine species of human. That we alone remain is testament to our aggressive, expansionist success, wiping out many species on our march to dominance, from dodos and all of Australia's megafauna, to the recent ivory-billed woodpecker and splendid poison frog (the first two examples when I asked Google which species have gone extinct recently). We are uniquely dangerous.

But our success over the other *Homo* species was also down to our superior skills of communication and community. Yes, we wreck everything, but we are also well suited to fixing problems, if only we choose to do so. We need now to tell the stories that will ignite everybody to care about the perilous state of nature and the impact its collapse is having on people across the world. And then we need our local, national and international communities to work together to turn that around. Will we choose to balance our remorseless progress with concern and empathy?

Until recently, this square had been empty and quiet, but now it was being swallowed by a sprawling new town. As I looked over the emerging streets, I struggled to take it all in. How does this happen? How can a town spring from the earth where my map still implied countryside?

Ignoring that I live in a house, buy things in shops and use roads every day, I'd rather see countryside than towns. But those facts are, of course, impossible to ignore. We all depend upon the concretification of the landscape for our modern, connected lives, and we all live on streets that were once meadow or forest.

On my shed wall I have a map from 1884 titled 'London Before The Houses'. It shows green woodland and marshy floodplains, with rivers that have now disappeared underground or into drainage channels: Bridge Creek, the Effra, the Fleet, West Bourne. The map helps me to imagine how things used to be, and the emergence of this new town reminded me to be mindful of the impact we make on nature.

Look out of the window. All the buildings you see were once not there. Everything is imposed on a formerly wild land. We need towns, of course, but it is worth remaining cognisant of their cost.

I set off along a footpath that was squeezed between temporary barriers to keep the public safely away from the building site. Hawthorn flowers, those darling buds of May, pushed through the railings, their froth of white blossom exuberant and full of life. Shakespeare's sonnet proclaimed,

'Rough winds do shake the darling buds of May,
And summer's lease hath all too short a date.'

Today's rough winds were swirling up clouds of dust from the building site. Even if we cover a grid square in houses, it is still part of the wild universe. Wind is simply air moving around, owing to the sun heating the earth unevenly. It blew explorers across oceans, mingles our pollution with our neighbours',* disperses seeds, whips sandstorms from African deserts to our northern skies, and dries our washing on the line. Invisible and untamed, wind scours and carves our landscapes. Harness some of that invisible power and we can generate virtually limitless clean electricity.

Winds around the world are given evocative local names such as Bora, Chinook, Haboob, Harmattan, Mistral and Scirocco. The only named wind in Britain is the Helm Wind, a strong north-easterly that hits the slopes of Cross Fell in Cumbria. I always found wind unsettling when travelling in remote environments on my own. I remember the few trees on the Patagonian steppe were all permanently bowed by the wind's relentless onslaught. But back home I like wind, along with every form of extreme weather. Windy days confirm that at heart I'm still a kid itching to run round the playground with my anorak lifted over my head like a sail.

CCTV cameras peered at me as I followed the metal fence around the perimeter of the development. While they looked at me, I looked at some split gill mushrooms growing on a fallen log and checked the time by blowing a dandelion clock. After flowering, the dandelion's petals dry and drop off, and a fluffy white ball of seeds develops. When the wind blows (perhaps a Bora, a Harmattan, or a breeze over this new town), those seeds soar into the air, carried by a white beard of bristles, known as a pappus, from the old Greek word for grandfather. An air bubble, known as a vortex ring and unique in nature, surrounds each seed while it spins through the air, stabilised by the geometry and design of the pappus. This increases the drag fourfold and slows its descent even more, allowing the seed to fly farther before landing and increasing its distribution.

**Musician Brian Eno tells a story of being invited to a rich celebrity's party in a scruffy part of town. The hostess held dear her opulent home and dismissed the rundown neighbourhood it was in as being 'outside'. He came to think of that approach as the 'Small Here', and contrasted it with the interconnected 'Big Here' world that we actually live in, where it is tempting but naive to lock a door behind us and pretend that what we do on our own small map does not impact the rest of the world out there.*

I continued down the narrow path that was overgrown with riotous green growth from the recent mild weather, then edged round an impenetrable mass of brambles and beneath a huge pylon, its cables all a-fizzing and a-crackling. I hate that sound and always imagine it is frazzling straight into my skull. An even more intimidating noise was the dual carriageway that the footpath now ran alongside. I was virtually on the hard shoulder and butted right up against the violent, roaring, stinking power and urgency of modern life. What would the local Neanderthals have made of all this? Who, or what, will be here in another 400,000 years from now?

Part of the new town was being built on the site of an old quarry. White cliffs dropped into a wide bowl that was already filled with completed buildings. I freewheeled down smooth new roads, past the blossom of a solitary cherry sapling, and followed my nose towards a large pond I had seen from my vantage point. I expected it to be out of bounds, danger, keep out, as usual, but I felt I ought to try.

And then there they were. Down by the water. Swifts! Hundreds of them screaming around the sky, winnowing the air, returned from their migration to Africa. They're back! I was elated to see them. Every year I celebrate my first sighting of a swift, as did poet Ted Hughes in his joyful poem 'Swifts'.

'Fifteenth of May. Cherry blossom. The swifts
Materialise at the tip of a long scream
Of needle. 'Look. They're back. Look.' And they're gone
On a steep
Controlled scream of skid
Round the house-end and away under the cherries...
They've made it again,
Which means the globe's still working, the Creation's
Still waking refreshed, our summer's
Still all to come.'

'They've made it again, which means the globe's still working': I love that line, and the optimism that nature can fill us with. It was ironic that I'd found the swifts on a grid square where I assumed the story was going to be about the removal of nature. Instead, I discovered a picturesque pond bordered by reedbeds, and birch trees rewilding the quarry walls with a steep greenness that felt more like the Austrian Alps than English suburbia. Crowning everything was the majesty of the swifts.

Swifts are right up there among my favourite wild creatures, and they score extra for heralding the return of summer evenings. I envy swifts, for they exist in perpetual summer, chasing warm days around the globe. They live in the heavens, ancient birds that have soared and swooped since they watched continents take form and mammals evolve. Today they hurtle around our towns' chimney tops, exuding joy and life itself. Swifts belong nowhere and everywhere, soaring nomads surfing on the winds of heaven.

The swift is Britain's fastest bird in level flight. It can hit speeds of 60mph and blast through 500 miles in a day, racking up a million miles or more in a life on the wing. When they leave their nests for the first time, young swifts do not touch earth again for a couple of years, until they return to breed in the place where they were born. They spend almost their entire life in the air, feeding, migrating, sleeping and mating on the wing. At dusk they climb to 10,000 feet, the soaring evening vesper flight, and drift half asleep through the dark skies.

I enjoyed watching the dramatic antics of the swifts as they barrelled above me like boisterous delinquents. In damp weather like this they swoop low to feed over ponds and lakes, but on fair days they gorge on flying insects high overhead. Each day they gather up to 10,000 aphids, midges and other insects, a hectic lifestyle that is now threatened by pesticide use and habitat loss.

Swifts visit us for the summer from Africa, navigating through celestial cues, magnetic fields and mysterious magic to nest in our eaves and share our buildings. We have lived alongside each other for millennia, but will not for much longer. Modern building designs are not conducive to nesting, and swifts are now in genuine danger of vanishing from Britain. We have lost more than half of our swifts in the past twenty-five years, placing them on the Red List of species threatened with local extinction. This is heartbreaking for a special bird that separated taxonomically from other birds way back when they flew among the last of the Tyrannosaurs. For swifts to be lost on our generation's watch would be shameful, but we do also hold it in our powers to be their saviours.

Inspired by the swifts' wild and enviable freedom, I hopped over a low fence and cycled round the back of the lake. It was ridiculous how guilty I felt about this 'trespass', though I was doing nothing more sinister than riding my bike along a track in an old quarry. It has become ingrained in me to feel uneasy every time I do a gentle, harmless thing such as this.

Behind the rustling reedbeds, a thicket of young trees and sea buckthorn flourished, a tucked-away corner testament to the wonders of self-regeneration.

It had been a surprise to find one of the quietest, wildest corners of my map on a noisy building site of a grid square. I would have gone for a quick swim, but was sure that would cause someone wearing hi-vis to yell at me. Instead, I just enjoyed sitting by the lake for a while, out of sight of anyone, watching wild swifts rejoicing overhead.

They've made it again, which means the globe's still working.

JUNE

FLOW

'If you know one landscape well, you will look at all other landscapes differently. And if you learn to love one place, sometimes you can also learn to love another.'

Anne Michaels

I dug out a pair of shorts to welcome in June. My legs shone alabaster white, brighter than the day's glorious sun. The lightness I felt inside made me aware of how sluggish I had been throughout the dark half of the year. Today, though, I was alert and enthusiastic. Even better, a chalk stream kissed the corner of today's grid square. So I began there, with the banks shaded by overhanging willow trees and lined with pink foxgloves, and with the clear water burbling. Trout nosed into the current beneath an arched brick bridge with an inscription saying it had been rebuilt in 1773. While the fish were free to swim, a 'Private Property' sign chained across the river prohibited curious explorers from enjoying the stream.

Chalk streams like this formed when the ice age receded ten millennia ago, shaping the distinctive gravel beds that still determine their ecology today. They bubble up from deep springs that run cold and clear thanks to the slow filtration of the rock. There are only a couple of hundred true chalk streams in the world, and 85 percent of them are in England, so they are perhaps our most important contribution to nature.

Chalk streams are beautiful and irreplaceable, yet we take no care of them. Our rivers are being destroyed by pesticides and run-off from industrial farming, and also by water companies dumping sewage or sucking them dry in response to political pressure to prioritise cheap water, and their shareholders' thirst for dividends. Every river in England is polluted beyond legal limits. Raw sewage was discharged into waterways 300,000 times in 2022, and 39 million tonnes of sewage floated down the Thames from London. Hotter summers have increased the number of life-smothering algal blooms, 10 percent of our freshwater species are threatened with extinction, and environmental protection agencies have been relentlessly defunded.

My mind boggles at the executives claiming enormous bonuses while consistently missing their pollution targets, and at the intelligent, educated individuals in Whitehall offices who've decided this is a legacy they are content to live with.

On the riverbank, a Metasequoia tree had been planted by 'a friend of the village and enthusiastic gardener on his hundredth birthday'. A plaque read, 'The *Metasequoia glyptostroboides* is the living relic of a fossil genus which thrived 125 million years ago. Discovered in China in 1941, its seeds were first germinated in the UK in 1948.' I am always intrigued by anything not related to warfare that happened during the Second World War, that some people were going about lives undistracted by fighting or fascism.

One of my favourite things about this year was how intrigued I had become by everything. Annie Dillard, that perennial enthusiast, wrote that 'if you cultivate a healthy poverty and simplicity, so that finding a penny will literally make your day, then, since the world is in fact planted in pennies, you have with your poverty bought a lifetime of days. It is that simple. What you see is what you get.'

Over the road from the fossil tree, a corn mill had been converted into prosperous homes. One had a beware-of-the-dog sign outside, '*Cave Canem*', a replica of the mosaic from the House of the Tragic Poet buried in Pompeii by the eruption of Mount Vesuvius in AD 79. In the same year, General Agricola attacked Scotland, perhaps using troops who marched up the Roman road that crossed my map. This

project had sent me ricocheting backwards and forwards through history much more than I had anticipated.

I pedalled past the old jail house, dated 1602, a cute white clapboard cottage that looked as though a small child had been the architect. The jail was part of a street of quirky old homes jammed up against the brutalist noise and speed of a motorway. The disparate pace of life spanning the centuries was discordant.

The rush of cars echoed off the concrete as I walked under the massive motorway bridge. Pigeons flew in from the sunshine to roost in the gloom. The construction effort of this bridge alone must have been monumental. It weighed thousands of tonnes and would have cost millions of pounds. There are 247,000 miles of tarmac roads in Britain today. We take that for granted, but if you compare that number with our 150,000 miles of footpaths, it gives an idea of how much life has changed since the village jail was last full of highwaymen.

Beyond the motorway, I headed up a path on the edge of a hillside meadow. Half the field had already been mown and gathered to be stored for animal fodder in the winter. Another sweet-smelling section had been cut and left to dry, and the remaining part of the field still awaited mowing. The grass was knee high and golden.

I enjoyed climbing the hill, sticking to my rule of not looking around until I reached the top, to ensure I got the full enjoyment of seeing the view for the first time in all its magnificence. It was a hot day, the first that really felt like summer. I looked forward to wombling around the wood that spanned the northern half of the grid square. It had been formally opened by Sir David Attenborough, no less. An information board promised it was 'a fascinating area of old woodland rich in trees, wildflowers and animals'. The sun was shining, I was in a wood on top of a hill: this was everything that makes me content.

Unfortunately, the wood turned out to be quite depressing, for most of it had been felled in a clear-cutting operation that went far beyond the usual levels of coppicing. There were piles of timber everywhere. It was hard to see the wood for the lack of the trees. Most of the sweet chestnut trees had been removed, and all that remained were some sorry-looking saplings and a couple of muddy, silted-up ponds. The motorways on two sides of the wood were louder than the feeble birdsong that remained.

All this destruction was caused by an invasion of the Oriental chestnut gall wasp. The wood had been cleared to limit its spread. I was relieved that it had been done to protect the wood's future, rather than as a permanent demolition job. We've had sweet chestnut trees in Britain for well over a thousand years (unreliable anecdotes claim the Romans imported them). For me, that feels long enough for the tree to count as an honorary native to our countryside. I'm content to embrace a slowly evolving landscape, like a stream's meandering route changes, providing it is not being diminished.

Purists, on the other hand, consider native trees to be only those that arrived without human input after the last ice age. In which case, we have just thirty-two natives in Britain. These are ash, aspen, bay willow, beech, bird cherry, black poplar, box, common alder, common juniper, crab apple, crack willow, downy birch, field maple, hawthorn, hazel, holly, hornbeam, large-leaved lime, midland thorn, pedunculate oak, rowan, sallow, Scots pine, sessile oak, silver birch, small-leaved lime, white willow, whitebeam, wild cherry, wild service tree, wych elm and yew.

I left the remains of the denuded wood and continued down a footpath frothing with cow parsley, six feet high and rising. I like its fresh heads of tiny white flowers and their deep smell of summer days. A pair of blackcaps sang their hearts out in a hedge, flitting among the branches with their shorter, less full version of a blackbird's chorus. The warbler's delicate song has earned it the nickname of 'northern nightingale'. I would never have heeded a blackcap before I began exploring this map, nor taken the time to pause and identify a bird by its song.

The sky shimmered over a landscape of trees, fields, villages and motorways. The river valley curved away into the distance, following the flow up towards the chalk stream's source in the hills beyond the boundary of my map. I no longer felt trapped on a claustrophobic, restraining map. This was my empire, as far as the eye could see. I felt that until this year I had lived on this map, but never really lived in it.

I felt myself drawn towards the neighbouring grid square, and from there on to the next one and the next one, onwards beyond the borders of my map, like Kipling's 'Explorer', for whom a voice whispered,

'Something hidden. Go and find it. Go and look behind the Ranges –
Something lost behind the Ranges. Lost and waiting for you. Go.'

But I knew I needn't be greedy, that I should wait until the time came to investigate those distant squares. There was so much to explore, but I had all the time in the world to do it.

ECLIPSE

'Will it not be employment enough to watch the progress of the seasons?'

Henry David Thoreau

The map promised waterfalls. I was not expecting the 979 metres of Venezuela's Angel Falls (named after the American explorer and pilot Jimmy Angel, whose plane crashed on Auyán-Tepuí in 1937), the volume of Inga Falls in the DRC (more than 46 million litres per second), or even the Denmark Strait cataract (an undersea waterfall plummeting unseen for 3,500 metres beneath the Atlantic Ocean). But the word 'waterfall' was not something I had expected to see annotated on my suburban lowland map, so I was excited to investigate.

My heart sank when I saw that the stream ran straight across a golf course. Golf courses are like a certain type of model. At first glance, your eyes light up at the swathes of undulating lushness. But your passion quickly plummets at the emptiness you find, the lack of nature beneath an artificial, preened veneer. The golf course did not bode well for my waterfalls.

Sure enough, the stream had been corralled into culverts and tunnels across the golf course and there was no sign of a waterfall. A second one was marked farther along in a wood. I waded up the water through thick undergrowth to find it, but the stream passed into a private garden so I didn't proceed. I could see that this waterfall had also been tamed, diverted into a concrete channel that fell forlornly a foot or two. I was disappointed, but not surprised. I would not be finding waterfalls on my forays after all.

At least I had enjoyed walking up the ankle-deep stream on this warm June day. Blue dragonflies shimmered around the water, although later research taught me that they were, in fact, damselflies, creatures I never even knew existed. Specifically, they were beautiful demoiselles, which are, well, beautiful. Male beautiful demoiselles are a gorgeous metallic turquoise colour, which they flaunt in elaborate flying dances, while the females have green bodies and brown wings. You might mistake them for dragonflies, as I did, but once you know that damselflies rest with their wings closed and dragonflies rest with them open, the distinction is easy.

When you see a path going into the woods, the sensible thing to do is to take it. It could make all the difference. I squeezed through a gap in a hedge, probably too large to be a smeuse, but the word is so superior to 'gap' that I'm going to squeeze it in here. I pushed into a meadow through thick ferns that gleamed in the sunlight with a fluorescent green glow. They are ancient plants that dominated the earth for hundreds of millions of years before flowers came on the scene. Dinosaurs roamed through spectacular ferns up to one hundred feet tall.

Meadows feel vibrantly alive in the summer, filled with colour and the buzz of insects. But once again, I'd arrived in today's square in a stupid mad rush to tick everything off. So I forced myself to slow down by putting the Seek app to work in the field of knee-high grasses.

The first plant the app identified for me was common sorrel, with edible lemony-flavour leaves and local names that include cuckoo's meat and butter and eggs. It prompted me to WhatsApp an old friend who had named his daughter Sorrel. I hadn't seen him in years, and in very different circumstances. Here I was, trying to slow down and remind myself that there was a world and a heaven in a tiny wildflower meadow in England, but also pinging a message thousands of miles across continents through the ether. My friend replied in seconds, pleased to be back in contact, and sending me photos of his Sorrel in the magnificent mountains where they now live.

There is wildness and beauty here; there is wildness and beauty on the other side of the world. Which is better? Perhaps the one you are in right now, so long as you're not just yearning for the other? That was my goal for this year, to find beauty in the here and now rather than spending my days longing for faraway places.

I was pleased when my map-reading skills pinned down an inconspicuous footpath on the far side of the meadow. We are lucky to have excellent maps in Britain,* and I appreciated being taught to interpret them well. There is a satisfying utility in being able to transpose those lines and symbols into a picture inside your head of what the land will be like.

The footpath I'd found was clearly not popular, for it was overgrown with nettles. June is a time of maximum daylight and maximum growth, the very opposite of the dark days of December. I braced myself, then launched myself in shorts and T-shirt into the stinger attack. I have a wary respect of nettles from a childhood dread of their stings. The Romans first brought nettles to Britain, combatting our miserable climate by flogging themselves with their stinging stems to generate heat and improve circulation.

A nettle's sting is like a hypodermic needle, injecting a venomous cocktail of histamine, acetylcholine and 5-hydroxytryptamine. This prickly threat, however, also makes them a haven for more than forty species of invertebrates who nestle between the stinging hairs. Such a bounty of prey in turn attracts amphibians, hedgehogs and birds. A patch of nettles left alone at the foot of your garden is a great hideout for nature.

Once I was heroically through the nettles (with much squealing), I channelled my inner Bear Grylls and picked some dock leaves to rub on the nettle stings. Everyone knows that dock leaves cure nettle stings. Except that they don't. Better to use *Plantago major*, aka the humble plantain. It's a pity I know so little about the wild plants all around us, and my inadvertent assumption is that animals are more interesting. This plant blindness was something I'd been trying to address this year.

The bluebells had vanished from the woods now, despite being

**Our maps date back to the intriguing Gough Map of 1360, one of the earliest to show Britain in a recognisable form.*

resplendent a few weeks ago. Only their stems remained. It was rare for me to register what went away as the seasons turned, to see what was not present, as I was usually so distracted by all the new stuff that appeared every week.

A baby squirrel scampered across my path, then ran up a tall horse chestnut tree whose large, candle-like flowers were in full bloom. In a couple of months, these would become conkers and be gathered by what, by then, would be a full-sized squirrel. Conkers contain the toxin aesculin which squirrels can only tolerate in small amounts. So one of its hoard may then germinate and grow into a new tree.

Continuing through the patchwork of farmland and horse paddocks that made up this corner of the map, I saw a small yacht, forgotten and forlorn, mouldering beneath a tarpaulin and marooned in a field far from a rising tide. I wondered what percentage of yachts ever put to sea, and how many days they spend under sail. With all the oceans of the world awaiting, my wanderer's heart went out in sympathy to the stranded boat and her unfulfilled dreams.

A handful of pretty stone cottages, hundreds of years old, clustered around an old manor house. Trellises of roses arched over the doorways. A sign gave the address of the manor's website, but visiting it redirected me to an eye-opening Asian porn site. Moving on from this idyllic, timeless hamlet (and pouting webcam girls), I followed a narrow path wedged up against the railway line (all hail to the nettle-strimmer who had cleared the way).

In a field of ragwort I came across a yard of lock-up garages, filled with cages of snarling dogs who intimidated me even from behind their bars. CCTV cameras monitored the space, a punchbag hung in the empty work yard, and the area around the unit was covered with knackered vans and piles of half-burnt rubbish. Whatever went on here was a million miles (and a few hundred metres) from the tranquil scene of respectability I had photographed up in the hamlet.

I didn't linger with my camera, but moved on into the next field, where my map promised a pond. But instead the entire field had been bulldozed in preparation for new houses and was covered in multi-coloured fragments of plastic. There was more plastic than pebbles in the field.

Mass production of plastics only began in earnest after the Second

World War, but it has spiralled so far out of control that we have now made 8.3 billion tonnes of it, mostly as disposable products. Three-quarters of all that plastic has already become waste and less than 10 per cent has been recycled. As it takes several centuries to break down, it all still exists in landfills or the oceans, where 8 million tonnes of plastic ends up each year.

Imagine fifteen shopping bags filled with plastic piled on every single metre of shoreline across the entire planet: that's how much plastic we dump in the oceans every year. At our current rate, by 2050, there will be more plastic in the sea than fish.

Meanwhile, the natural world battled on, miraculously and resolutely. Delicate blooms of elderflowers filled the hedgerows around the field filled with plastic. I thought about gathering some to make elderflower cocktail. I like the idea of foraged food using simple, local ingredients. But the recipes I found online were heavy on imported lemons and sugar, with only a few flowers from the local hedge chucked in for good measure. So I gave up on that plan and instead cheered the first green woodpecker I had seen on my map, a chunky fellow who bobbed and hopped over the ground. The green woodpecker uses its tongue to forage for ants. It is an astonishing organ, so long that it coils behind the skull, over the bird's eyes, and into its right nostril! It is one third the length of the bird's entire body. Cumbersome, perhaps, but magnificent for feasting on ants.

My phone alarm rang as I was crossing back over the golf course on a footpath that ran precariously across the driving range. I had set it to remind myself that a partial eclipse of the sun was due to begin. The last eclipse I'd seen was while driving from Las Vegas to Zion's magnificent cliffs and canyons. I was excited by today's homegrown affair.

I settled down on the outfield of a cricket pitch to lie back and look up at the sky, but the sunny morning had now clouded over. This was frustrating for my appointment with the heavens, but minor compared with the tribulations of French astronomer Guillaume Le Gentil.

Commissioned by the French Academy of Sciences to record Venus's transit of the sun, Le Gentil spent months sailing to India in 1761 to observe the event. Alas, the Indian territory of Pondicherry was under siege from English warships, so he was sent back to Mauritius and missed the transit. Rather than give up, Le Gentil stayed in Mauritius, making plans to study the next transit of Venus instead. In 1766, he set sail for Manila to observe the transit, but was accused of being a spy. He quickly fled to Pondicherry, hoping to catch it there instead. But on the day of the big event in 1769, the sky clouded over, and he saw nothing at all. Le Gentil took a deep breath and decided that enough was enough. He set off for home, a journey that in itself

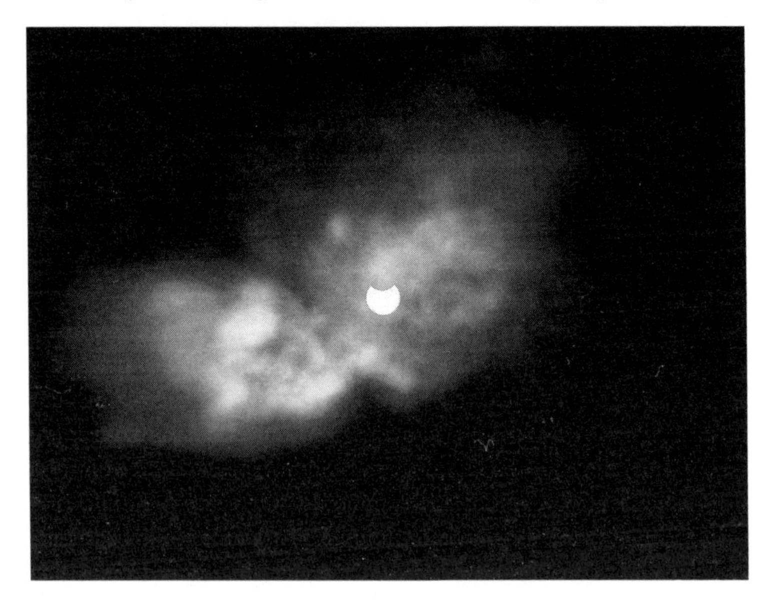

took more than a year to complete, and eventually made it back to France in 1771, a decade after first setting sail.

Having stared at clouds for all of two minutes, unable to see the stupid eclipse, I knew exactly how Le Gentil felt. Fortunately for me, my immense patience was rewarded, and the cloud suddenly thinned into perfect eclipse-watching conditions. There was just enough cloud cover to save me from frazzling my retina like some sort of moronic American President as I pointed my camera up at the sun in search of Instagram glory. The moon passed in front of the sun, munching a fair-sized bite from the bright disc. It was so exciting to watch. Lying on my back on a local cricket pitch, I marvelled at this reminder that there is a hell of a big universe out there beyond my map.

MEADOWS

'The answer must be, I think, that beauty and grace are performed whether or not we will or sense them. The least we can do is try to be there.'

Annie Dillard

I had a free morning and my latest grid square lay before me, beginning with the rare pleasure of a segregated cycle lane, safe from the busy road that sliced the square in half. I rode fast and free, blasting away the day's earlier frustrations of waiting on the phone for an hour to speak to my electricity provider. Free at last! (Me, not the electricity.) North of the road, wheat fields ripened in the heat. South of the road lay a 1940s housing estate. The noisy road was once an important Roman route, though it was already an ancient thoroughfare by the time they arrived. I can't begin to imagine what the traffic here will look like in another 2,000 years.

A row of houses had recently been built between the road and those wheat fields that had been forest back when the Romans carved through this land in the name of progress. The new-builds were extravagant expanses of glass and steel, with large gravel areas for parking multiple cars. Sparrows jostled noisily in pink rose bushes and petals fell among the squabbling. A placard in one garden campaigned against a 'green belt grab' that proposed to build 4,000 more homes around here. It summed up the difficulties of deciding where to build. This family was enjoying their new home but understandably didn't want all the neighbouring fields to be built on as well. I don't like the countryside being turned into towns, but I also want everyone to have a home. Answers on a postcard to your MP, please.

I turned down a farm track and passed a fly-tipped safe, a novel addition to the usual broken sofas, babies' highchairs, and bags of builders' rubbish. It was stolen, no doubt, but still unopened. I imagined the thieves' frustration as they dumped it in this field after failing to smash the door open. The land here was so flat that a hill to the north about a hundred metres high looked majestic. The summer verges between the rubbish were flush with meadow foxtail, ribwort plantain, cocksfoot, salsify and buttercups. Heat reflected up at me from the stony track. Larks sang overhead, sounding as pretty damn delighted with these sunny days as I felt.

Across the main road lay the sprawling estate. As I pedalled around the streets I stopped to watch a snail slither across a pavement. The top speed in the Guinness Gastropod Championship, held over a 33cm course in the O'Conor Don – London's oldest Irish pub – is a dizzying 9.9 metres per hour, racked up by a snail called Archie who slithered to victory after the starter's signal of, 'Ready, Steady, Slow!'

Does a snail experience the world differently from me? Does it feel it is missing out by moving so slowly and seeing so little? Or is it mindful of all that it encounters and deems that local richness to be more than enough? Or is it merely an invertebrate with a set of ganglia and no brain with which to dredge up laborious metaphors?

I wasn't sure where the snail was going, or what it hoped to eat once it got there. At first glance, the only plants on the estate were the weeds growing through the cracks of a basketball court whose hoop had been ripped from the backboard.

I pedalled past a primary school and the smell from the kitchen triggered a sensory flashback for me, an involuntary memory in search of lost time, back to school dinners of powdered mashed potato and compartmentalised plastic trays. The pub next door was offering three shots for £5 (prompting student flashbacks) at the end of a row of off-licences and takeaways.

Fast food is nothing new, despite the modern health crisis it is contributing to. The story of the humble chip, for example, dates back to the 17th century, with both Belgium and France claiming ownership. Chips may actually have been invented as an alternative to fried fish, rather than as its perfect partner. One harsh winter, when the River Meuse in Belgium froze over and no fish could be caught, resourceful cooks in Namur began slicing potatoes into fish shapes and frying them as a substitute.

It was Jewish refugees from Iberia who brought fried fish to Britain, selling it from trays hung around their necks. Bloody foreigners, coming to our country, working hard, bringing tasty recipes and saving us from our pottage and gruel.

In *Oliver Twist*, Dickens described a scene not too dissimilar from today's high street, with 'its barber, its coffee-shop, its beer-shop, and its fried-fish warehouse' where the fish was served with bread or jacket potatoes.

By 1863, John Lees was selling fish and chips from a wooden hut in Mossley market in Lancashire. Southerners like to claim that Joseph Malin's East London fish and chip shop was already in business by 1860. Beyond dispute is that the concept galloped across Victorian Britain, and by the 1920s there were more than 35,000 fish and chip shops. There are actually fewer takeaways today, though fish and chips has been overtaken by burgers, fried chicken, pizza, Indian and Chinese food.

A thin strip of tangled weeds ran between a wooden fence and a footpath dotted with fast-food rubbish. I stopped to look more closely and found barley, dandelions, nipplewort, common mallow humming with pollen-dusted bumblebees, and many more plants I didn't know. I followed the path to see where it led and was thrilled to discover a small wild common and nature reserve tucked up behind the housing estate. It was a delightful surprise and a real treat.

I walked for an hour around the chalk downland without seeing another person among the long grass, scrub, and thickets resulting from old hedges spreading. Blackcaps and blackbirds sang their summer anthems from the hawthorns. An information board boasted that twenty-nine bird species lived here. I wasn't sure I could even name twenty-nine birds.

With only 3 percent of our wildflower meadows remaining, I felt so fortunate to have found this one on my map, a wonderful habitat supporting a web of insects, animals and birds. In the past century we have destroyed an area of wild meadows equivalent to one and a half times the size of Wales. So I had been pleased to read about the success

of this year's No Mow May campaign, with councils in London mowing one-third less grass than usual, resulting in messier, better spaces for wildlife.*

My Seek app ticked off tufted vetch and hairy vetch, orchard grass, common soft brome and wild clover, the shimmering hearts of quaking grass, purple scabious, ribwort plantain, lilac knapweed, ragged robin and autumn hawkbit. What magical names! The colours and textures of all these grasses and flowers waved in the sunlit breeze, as they have done for millions of years.

This was a mere scrap of wildness, only about the size of twenty football pitches, but it was a tiny haven that had been left alone since the estate was built. Apart from glimpses of houses between the trees and some kids' graffiti on the sign boards ('Jesse West is going to get his shit rocked by Eliza Valicy'), I could have been walking through a much older and wilder landscape. This was how much of Britain used to look. And it had all reestablished itself, from managed farmland, in just a few decades. Once we start farming and eating more thoughtfully, we can make so many more spaces like this that benefit both people and nature.

**Consider also: Let it Bloom June and Knee High July.*

SOLSTICE

'This is June, the month of grass and leaves... I feel a little flustered in my thoughts, as if I might be too late. Each season is but an infinitesimal point. It no longer comes than it is gone.'

Henry David Thoreau

I sheltered beneath a large field maple tree, reframing my attitude to rain. Parking the grumbles and persuading myself instead how gleaming clean all the trees looked. Appreciating the gun-barrel-granite skies. Remembering that a day in the rain is better than a day in the office. That kind of thing.

One of my favourite smells is the air after a storm, the earthy scent of petrichor, from the Greek words *petros* (stone) and *ichor* (the blood of the gods). We tend to think that our sense of smell is something to be sniffed at compared with the animal world's, but we are astonishingly adept at detecting geosmin, the chemical released by dead microbes that is responsible for the heady smells of petrichor and pools of water. We can smell geosmin at a level of five parts per trillion – that's thousands of times more sensitive than sharks are to the scent of blood. We may be so sensitive to it because detecting water on the savannah where we evolved was a vital evolutionary advantage.

There was no shortage of water for me to sniff today beyond the boughs of my tree umbrella. Rain bucketed from the sky and sluiced off the fields. But, motivated now, I ducked down a narrow holloway shiny with raindrops and spiky with nettles.

So often on the obscure footpaths I took on these grid squares, I had the thought that I must be the first person ever to go this way. Who else would have taken this short path from nowhere much to nowhere else? But, of course, each footpath exists because they have been walked by unseen footfalls over hundreds of years, originally to get from here to there, and these days to walk the dog while checking emails and listening to music. You can't easily make your own footpath. They are a collaborative effort. Without all of us unseen partners doing our bit, paths will not survive the onslaught of brambles, building or ploughing. I usually assumed that I saw nobody because I went out during the working week, but it might also be because I went out in the pouring rain.

So I had the holloway to myself, though its very existence was proof of footsteps beyond number passing this way. Holloways sometimes feel more like tunnels than tracks, with tree branches curving overhead and the ground worn away by the tread of traffic over centuries, assisted by water erosion. Tangles of tree roots gaped from the banks. Raindrops pearled on purple foxgloves. Fungi grew in the damp leaf litter, small shy huddles of grey-capped mushrooms. I spotted a cluster of King Alfred's cakes, a round black fungus that grows on rotting wood, particularly beech and ash. It darkens with age and lives on the decaying deadwood for years. When dried, it makes useful tinder for lighting fires, burning like charcoal, but with pungent smoke.

The fungus gets its name from King Alfred's famously poor baking skills when he was on the run back in the year 878. Alfred took refuge from the Vikings in a peasant's home in Somerset. She allegedly asked him to keep an eye on her cakes (small loaves) baking by the fire. Domesticity rarely being a forte of royalty, the king let the cakes burn, earning himself a scolding for his scalding.

I had come out today partly through habit, despite the rain, but mostly because it was the summer solstice. Midsummer and the longest day. Highlights of my year include Christmas Day, my birthday, and 16 June – the date I left school. And 21 June is right up there amongst

them. The solstice doesn't always land on that day, but to aid my little brain this is the date I mark it, noting my shortest noonday shadow of the year.

The position of the sunrise varies significantly over the course of a year, from northeast to southeast. In the northern hemisphere, the June solstice marks the sun reaching its northern limit. Today was also the astronomical start of summer, though meteorologists mark the season as beginning on 1 June.

It was lashing down, but I knew from experience that solstice rain is not a rarity. For many years, I encouraged tribes of enthusiastic adventurers to sleep on their local hill for a summer solstice microadventure challenge, and we often seemed to get soaked.

I followed a path along the edge of a field sown with young corn. A melanistic rabbit nibbled away at the crop. These darker animals are a natural variant that persist despite being much more visible to foxes or weasels. The laws of heredity, which dimly remind me of school lessons about monks and pea plants, produce a certain number of black rabbits per population.

By now my bike's wheels were so clogged with mud that I had to resort to carrying it until I reached a road. So much for midsummer! Jackdaws chattered in the trees as I grunted along. The jackdaw call is a familiar hard 'tchack' sound that gives the bird its name. It always reminds me of midwinter dusks and the smell of coal fires. Unlike their noisy cousins, the rooks, jackdaws get by with brief yet meaningful conversations. They pair for life, share food, and are smart enough to recognise human faces and communicate using their eyes, as we do.

While I was watching the jackdaws, I spotted a crow flying away clutching a baby bird from a treetop nest. Daily life in nature revolves around such moments of life and death, hunger and loss. If I had passed by a few seconds earlier or later, I would have missed the commotion. There is so much drama we never witness.

Spending time in woodland always cheers me up, so I was pleased to head into some more trees on the north of the grid square. The rain had stopped, but water continued to drip from the branches. I sat for a few minutes near a badger sett with tonnes of sticky earth heaped outside, just listening and watching. I must have been hard to spot in the green gloom, wearing a green raincoat, for a blackbird flew straight at me, then did an emergency last-minute left turn around a beech tree to avoid me, its wings beating audibly like a drum.

The warblers, tits and thrushes were in full voice. Birdsong begins extremely early, with the dawn chorus. It is a sign that the day has started and it's time to get moving, so birdsong also stimulates us cognitively. Scientists have studied the restorative benefits of birdsong for helping people to recover from stress and restoring focus.

The atmosphere in woods would be very different if birdsong was a jarring noise (like donkeys braying, for example). We are conditioned to find birdsong relaxing and reassuring because, over time, we learnt that when they were singing we were safe from predation. When the birds stopped singing, it alerted us to start worrying. A silent spring should fill us with fear.

The Seek app continued to teach me, as well as helping me to travel slowly and be alert. Today I learnt about the stinking iris with its orange fruits sheathed in pods. It is said to stink of rotting meat, but I had a cold so couldn't smell anything. A burst of yellow fungus also caught my eye on a dark, wet log, and I discovered it had the magnificent name of hairy curtain crust.

Down at the foot of the hill, clusters of red poppies stood out against the green hedgerow, their petals like crinkled tissue paper and covered with raindrop jewels. Poppies are the source of opium and morphine and have been used since ancient times for pain relief, recreation and for the sprinkly little seeds you get on fancy bread rolls. The flowers have been symbols of remembrance since they bloomed blood red and heartbreaking on the muddy, wrecked fields of the First World War. Canadian doctor John McCrae featured poppies in his poem 'In Flanders Fields', written after the battle of Ypres:

'In Flanders fields the poppies blow
Between the crosses, row on row,
That mark our place; and in the sky
The larks, still bravely singing, fly
Scarce heard amid the guns below.'

Bees feasted among the poppies on the purple flowers of spiky thistles, while mullein moth caterpillars shredded burdock leaves. The internet told me little about the caterpillars, except for various gardeners' strategies for destroying them. Personally, I admired their livery of black dots on yellow and white stripes, and the adult moth's ingenious camouflage that resembles a twig.

Most of my expertise about caterpillars comes from *The Very Hungry Caterpillar*, which has sold over 55 million copies around the world. Eric Carle made the book using collage layers of tissue paper to tell the story of a greedy caterpillar scoffing every child's dream feast. He hatched the idea while playing around with a hole punch and imagining a worm eating through a book.

His editor suggested a caterpillar would work better than a worm, resulting in millions of sales and almost as many indignant letters pointing out that caterpillars form chrysalises, not cocoons. I only hope that the many mistakes I've probably made in this book result in equally vast numbers of sales, if not so many complaints.

JULY

JADED

'Man is the most insane species. He worships an invisible God and destroys a visible Nature. Unaware that this Nature he's destroying is this God he's worshipping.'

Hubert Reeves

Each week I arrived in my grid square with little idea what might capture my interest, but an increased certainty that something would. As with all good exploration, there were hints and hopes about what I'd find, but each square also surprised me.

This meant that if I found a square underwhelming, with little to interest me, the responsibility was likely to be mine. Was how much I saw dependent on how much I looked? Some squares buoyed my mood, while others merely matched it. A boring square wasn't its fault; it was my fault. I knew that as I struggled lethargically round today's streets, but I also excused myself on the grounds of illness.

I had sweated and shivered through the night, unable to sleep. In the morning, I went to make myself some toast, but we'd run out of bread. I dragged myself to the shed to do some work, but after an ineffectual hour of pretending to write this book, I tried to salvage something useful from the day by fetching my camera and cycling out to investigate a grid square.

What I'd seen on the map before heading out was the total concretification of today's square. There was only one footpath, running for just 80 metres behind a row of houses, and a tiny scrap of thicket tucked between a fold of main road and a roundabout. The rest of the square kilometre was a man-made mosaic of streets, railways and roads. Washing lines and trampolines. Rows of terraced houses. But roses too, for my first surprise on arriving was how much greenery there was in this ordinary town.

Potholed alleyways had grass growing in the middle, and brambles and nettles sprouted around each garage topped with barbed wire. A cedar spread its boughs in front of a development of tiny retirement bungalows. The grass outside a tower block was full of clover and buttercups. Elder bushes had taken seed on the flat roofs of a row of lock-ups. Hawkweed flowers burst through the broken concrete in parking bays. The neglected gaps between semi-detached homes were filling with undergrowth (and mattresses, beer cans and plastic bottles). Almost 15,000 species live alongside us in urban areas, and 'nature goes on existing unofficially', with George Orwell's pleasing thought that none of it 'pays a halfpenny of rent'.

The embankments of the railway line looked as lush and impenetrable as a jungle. Collectively, embankments add up to thousands of miles of wild land and have become important corridors for nature. Railway tunnels and cuttings also offer habitats for animals and plants to thrive in.

Network Rail says they meet 'and where possible exceed the legal requirements when it comes to protecting species and enhancing their environment. Our project work can have an immediate impact on local biodiversity, but we're testing methods of giving back to the natural environment more than our work has taken – often known as a net positive approach.' This was heartening to read.

I tucked down an alley behind a hostel for homeless people and out onto a high street busy with traffic, fast food franchises, cafés, barbers, charity shops and convenience stores, including a Slovakian, Bulgarian and Lithuanian grocery store. These shops serviced the rows of terraced houses covering the grid square, built in stages over the past century. The oldest streets bristled with chimneypots, the intermediate ones, with no chimneys, had front gardens, and the newest developments had neither chimney nor garden, just paved space to park as many cars as possible.

'Stop Smoking, Start Vaping' urged a sign at the vape shop. Good for business, I guess, but the first half of the sentence on its own would have been better advice. I have never tried vaping, so was tempted to pop in and indulge my curiosity. But I have enough bad habits, so I just carried on, riding past an antiques shop offering 'Antiques Olde and Modern'. I was feeling weak again, so popped into a newsagent's and emerged with a homemade vegetable samosa and a Diet Coke to sort me out.

Despite being considered a classic Indian snack, the samosa's origins are complex, like many tales of immigration and assimilation. They weave back at least as far as a 10th century Persian text mentioning the *sanbosag*, or 'triangular pastry'. Variations travelled and evolved along the trade routes between North Africa and East Asia. The samosa eventually reached India via Middle Eastern cooks hired to work in the kitchens of Muslim nobles. A 14th-century account of the court of the Sultan of Delhi describes the '*sanbosag*, a small pie stuffed with minced meat, almonds, pistachio, walnuts and spices'.

Diet Coke's history is less storied, and its ingredients are more mysterious than the humble samosa's. Ordinary Coke contains a ludicrous seven teaspoons of sugar in each can, so I had switched my bad habit to Diet Coke. Coca-Cola tries to minimise the fakeness of the sweetener aspartame by saying it's made from building blocks of protein found in everyday foods like meat, fish and eggs. That sounds nicer than describing a methyl ester of the aspartic acid/phenylalanine dipeptide. Either way, I know I'd be better off heeding Cristiano Ronaldo's viral advice to drink *água* rather than Coke. Anything artificially sweetened with something weird made in a lab can't be good for you. Heed Rule 19 from food writer Michael Pollan's excellent guidelines for eating well.

'If it's a plant, eat it. It was made in a plant, don't.'

I also like the most famous of his recommendations.

'Eat food. Not too much. Mostly plants.'

The cheerful chatter and laughter of playtime in a primary school carried down the length of the street. Matching the children for exuberance was a quarrel of sparrows in a privet hedge (another perfect term of venery). It was the only strip of hedge left on the road, so it had become a haven for the birds. Sparrow numbers are plummeting, with London losing 60 percent between 1994 and 2004. Culprits range from cats to pollution, but an inexorable paving over of their habitat certainly doesn't help. It's a far cry from the book *Nature Near London*, which marvelled back in 1883 that 'sparrows crowd every hedge and field, their numbers are incredible'.

As I pedalled around slowly, still feeling sorry for myself, I shared the streets with Amazon delivery vans, Deliveroo mopeds doing lunch runs, and some young mums out with their toddlers. A man leant on an upstairs windowsill, drumming his fingers, gazing out, but whatever he was seeing was inside his mind, not down here on the street. Two old men on a street corner chatted about the England cricket team. A couple of roads down, I passed two other men discussing the local non-league football team. Someone was hoovering their car. A postman knocked on a door then turned away when nobody answered. A window cleaner reached up to the first floor with a long mop. An old lady stood and watched him.

What makes a grid square interesting? Is it everything that lives there, or at least the things you notice? Is it what makes you curious,

sparking connections and ideas? Or is it an intangible sensation of how a place makes you feel, its sense of place? Within this square, people were living and dying. Loving and crying. It was busy with people driving and walking, working and resting, eating and drinking, sleeping and chatting. I might well have passed one of the ten murderers we each walk by in our lives or missed a soulmate for life by minutes.

'None of us thought of the others we would never meet or how our lives would all contain this hour,' wrote Philip Larkin. Every map square is a miniature of a larger world, and perhaps each one has anything you choose to find in it.

As J.A. Baker wrote in *The Peregrine*, 'the hardest thing of all to see is what is really there'.

And Thoreau summed the challenge up well, of course, 'it's not what you look at that matters, it's what you see'.

How you look, what you see, and the way all this makes you feel: a single map and the best of all possible worlds.

HOVERING

'At present, in this vicinity, the best part of the land is not private property; the landscape is not owned, and the walker enjoys comparative freedom. But possibly the day will come when it will be partitioned off into so-called pleasure-grounds, in which a few will take a narrow and exclusive pleasure only, when fences shall be multiplied, and man-traps and other engines invented to confine men to the public road, and walking over the surface of God's earth shall be construed to mean trespassing on some gentleman's grounds.'

Henry David Thoreau

This kingdom of mine might cover only twenty kilometres squared, but it seemed at times to span a thousand worlds. From winter to summer, welcoming smiles to grumpy shouts, and from last week's jaded streets to this long grass, busy with butterflies, where I lay on my back, alone and undisturbed, and enjoyed the warm sun on my face.

Down in the distance I could see the city's gleaming towers, shimmering in the midsummer haze. I lay still for a while, listening, hovering above myself in my mind's eye, allowing myself to settle into the grid square and its vibe. I heard birdsong and the hum of a motorway. 'The language of birds is very ancient,' wrote Gilbert White in a letter. 'Little is said, but much is meant and understood.'

Shaking the warm lethargy from my limbs, I walked over a couple of fields and found myself in one of the most beautiful landscapes of my map. It was a postcard snapshot of rural England, all soft undulating curves and hues of green. The view was funnelled down a steep valley whose flanks were a voluptuous expanse of wild meadow, overflowing with grasses, buttercups and tall spears of blue viper's-bugloss.

I took off my shirt as I climbed the hillside. This was the first time I'd unleashed my spectacularly white torso this year. I'm seriously lacking in melanin (unlike the rabbit I saw recently) so am always the pastiest fellow on the beach. The sun hung hot overhead in the still air. Sweat ran down into my eyes. Summer had definitely arrived.

Today was St Swithin's Day. According to legend, whatever the weather is like on his day remains constant for the next forty days and nights.

'St Swithin's Day, if it does rain
Full forty days, it will remain,
St Swithin's Day, if it be fair
For forty days, t'will rain no more.'

One of the criteria for becoming a Christian saint is performing a 'verifiable' miracle, and Swithin's credentials appear rather modest in this department. His solitary miracle involved fixing an old lady's broken eggs. When Swithin, the Bishop of Winchester, died in the year 862, he declined the usual pomp of a prominent burial in the cathedral. He asked instead to be buried outside in a simple lot 'where the sweet rain of heaven may fall upon my grave'. However, a century later, his bones were moved to an ornate indoor shrine so that pilgrims could benefit from his miraculous egg-mending powers.

The saint sent down a great storm in retribution for this, and it apparently rained for forty days and nights. A meteorological legend was born. However, since records began, there has never been forty days of consistent weather following St Swithin's Day. Perhaps the French, who look out for rain on St Gervais's Day, or the Germans' Seven Sleepers Day, have more luck in predicting the weather?

Such musings distracted me as I sweated and puffed my way up to the top of the hill. Chalk downlands like these are Britain's mini version of a tropical rainforest, in that they are unique habitats packed with life. As many as forty species of plant can live in a single square metre, and many grow in few other places. Conditions are harsh on the lime-rich, thin soil. Because the ground heats up and dries out quickly, the usual grasses do not dominate, particularly if combined with occasional mob grazing from livestock. This gives an opportunity for rarer species to thrive, including several beautiful orchids.

Looking north now, I could see right off the top of my map and on towards other people's maps and lives and stories. Bees busied themselves among the clover, and red admiral butterflies flittered in the fierce sunshine. I picked up my pace and sought the shade of the woods ahead.

This may be a heretical suggestion after all the muddy moaning months of slopping through soggy woods in winter rain, but summer could be the worst season to walk in woodland, the most dull and lifeless. There is little birdsong, brambles block many of the views and smaller paths, and the woods feel sombre, with so little light breaking through the dense overhead canopy.

Both birdsong and bird numbers seem to decrease in midsummer. Once birds have completed their breeding they no longer need to sing to defend territories. They also moult into new feathers, during which time they are less able to escape predators, so hide away as much

as possible.

The wood boundary was garlanded with trespass signs, which I ignored, but I couldn't ignore the nagging sense that I was unwelcome. Although I didn't see anyone all afternoon, the feeling of being watched followed me across the square. In *The Growing Pains of Adrian Mole*, Adrian wrote, 'Saturday, June 19. Nigel and I went for a bike ride today. We set out to look for a wild piece of countryside so that we could get back to nature and stuff. We pedalled for miles but all the woods and fields were guarded by barbed wire and 'KEEP OUT' notices so we could only get near to nature.'

Back in open country, a kestrel flew overhead. They are one of my favourite birds, 200 grams of creamy, mottled aerial assassin. Kestrels' remarkable eyesight helps them to catch as many as eight voles per day. They can also see ultraviolet light, which helps them to spot urine-tagged vole trails that glow under UV light. The birds have learned to monitor farm machinery too, waiting for tractors to unearth tasty morsels, for they can spot a beetle from fifty metres.

Kestrels are masters of hovering, which they do by facing into the wind and letting it lift and hold them. Their feathers are designed to reduce turbulence and avoid stalling, helping them to hang in the sky like a kite, their heads held perfectly still, ready to plummet and hunt.

Marginally less graceful than a kestrel, but not by much, was the distinctive outline of a Spitfire fighter plane that flew overhead with its characteristic throaty roar. The Spitfire is the most famous aircraft from the Second World War, treasured by both military buffs and misty-eyed nostalgics.

In some of the most pivotal weeks of the war, the Luftwaffe threw 2,600 aircraft against the RAF, who had just 640 fighters. Hermann Goering predicted victory within days, but Britain went into a frenzy of fighter-plane production. The Spitfire's speed, manoeuvrability and firepower helped to win the Battle of Britain and turn it into a British icon.

In its early iterations, the Spitfire was often outmanoeuvred in duels with German aircraft, resulting in many pilots getting shot down and wounded. Hundreds of these young men's appalling burns were treated at Queen Victoria Hospital in East Sussex by Dr Archibald McIndoe, using experimental plastic surgery techniques.

The patients bonded and formed a drinking and social society known as the Guinea Pig Club. McIndoe said that it was 'the most exclusive Club in the world, but the entrance fee is something most men would not care to pay'. Members continued to meet for sixty years after the war, and always enjoyed singing their club anthem.

'We are McIndoe's army,
We are his Guinea Pigs.
With dermatomes and pedicles,
Glass eyes, false teeth and wigs.
And when we get our discharge
We'll shout with all our might:
"Per ardua ad astra,"
We'd rather drink than fight.'

Marginally less ferocious than a Spitfire, but not by much, was a pair of bright blue dragonflies (not damselflies) patrolling the hedgerow. Both the aquatic nymphs and the flying adults are aggressive hunters. They seemed far from water up here on this high plateau, but I'm sure they knew what they were doing.

Dragonflies (not the 1930s British twin-engined luxury touring biplane of that name) eat small insects such as mosquitoes (not the twin-engined, shoulder-winged, multirole combat aircraft from the Second World War) and midges (not the small, swept-wing subsonic light fighter aircraft prototype). The dragonfly was often depicted on Samurai helmets in Japan. It is a symbol of vigilance and focused effort, owing to its knack of moving in multiple directions while always facing forwards.

Beneath the Spitfire, the kestrel and the dragonfly, the meadows stretched out into a valley that shimmered in the heat. I could have sat there all day. But I had veered off the footpath to enjoy this view, and in the back of my mind I remembered that I'd been yelled at near here a few weeks ago. I felt a vague unease that someone might appear from somewhere and shout at me for something. This rubbed up against my sense of freedom, but also triggered my wimpy dislike of being told off.

I hovered between resistance and compliance, but in the end turned my back on the fine views and returned to the permitted route for the rest of my walk: the path of least resistance, if not the path heading to the sunlit uplands of open, responsible access for all.

CONNECTIONS

'What is this life if, full of care,
We have no time to stand and stare.'

W. H. Davies

A bonus round. A little something extra. Have a look at what you could have won…

I didn't go out today to explore a grid square as usual, but to see the squares between the squares. I'd found myself with the rare but joyous occurrence of a weekend afternoon all to myself, so decided to go for a bike ride to calm my nerves before the big football match in the evening. I wasn't playing and was merely preparing to take my seat in front of the TV with beer in hand and loud opinions galore. But the game was still all I could concentrate on.

I headed out after lunch to see how many of the grid squares that I'd visited I could link together in an afternoon. I would ride through as many as possible before I ran out of time, and then zoom home for kick-off. It would be interesting to take stock of all I'd seen so far.

Calculating the shortest route between multiple points is known as 'the travelling salesman problem'. It is notoriously difficult, being, apparently, an 'NP-hard problem in combinatorial optimisation, important in theoretical computer science and operations research'. I can't pretend to understand what that means, let alone know how to optimise an itinerary. I engaged my brain for approximately five minutes by plugging the grid squares I'd visited so far into Google Maps, and left the rest to its increasingly competent skill at finding adequate cycle routes. I didn't really mind where it took me. I just wanted to ride. Then I jumped on my bike and got going.

It felt strange to set off without a specific grid square in mind, and liberating not to be lugging a heavy camera or feeling duty bound to pay close heed to everything. I was just going for a burn-up. Once you have built the habit of noticing, however, I noticed that it is hard to stop. So I noticed (and appreciated) the smoothness of a stretch of new tarmac. I noticed a squashed hedgehog on the road into town, sorry to see it, insides and all, but also kind of pleased to see it, for dead hedgehogs imply live hedgehogs, and I had seen vanishingly few on my map over recent years.

I soon found myself on streets I didn't know, gliding through squares I hadn't been to before. I chose each week's square at random, so if I'd been sent to this grid square rather than that one, how would things have changed? I was sorry I had not travelled to both. At the very least, those changes would have altered the route of today's ride and given an alternative perspective.

I may not have seen, for example, a blonde toddler frown and hesitate to push her pink scooter through a muddy puddle. I wouldn't have heard her indignant tears when the wheels got dirty, or her muscular father consoling her in Polish, or enjoyed our grins of mutual understanding as he switched to English to say to me, 'Daddy's in trouble again.'

I would have missed the teenager running into bowl in a cricket match, glimpsed between gaps in a hedge as I pedalled past. The batsman stood back in his crease, defended, and there was no run.

I might not have learnt about the useful footpath that cut down behind the houses, with a colourful mural saying 'HAKUNA MATATA' above overflowing bins and a fly-tipped TV. The path led down to some barges on the river, marooned on grey mud until the tide lifted them again.

A very young bride and groom posed for wedding photographs

outside their front door, and I startled them with a joyful cheer as I whizzed past ringing my bell. All afternoon, through the tall heat, I saw so many things I would not have seen had my route taken me another way.

I zoomed along, savouring my freedom, enjoying the lightness of having nothing to do but pedal fast. Google was in charge of the navigation, and I felt no pressure to pause, take notes, or do anything but ride and enjoy myself. I took sloppy phone photos from the saddle, firing from the hip without even bothering to stop. As I passed from one suddenly familiar grid square to another via linking roads and areas that were still new to me, I was struck by what a multitude of worlds were tucked within the folds of my small map, like the founding myth of the ancient city of Carthage.

King Mattan of Tyre had a daughter named Dido, a nickname meaning 'wanderer'. Her brother, Pygmalion, cheated Dido out of her inheritance then killed her husband and claimed the throne. Dido escaped west with her supporters, landing in North Africa to found a new city. But King Hiarbas, the local leader, granted Dido only as much land as she could cover with an ox hide. Wily Dido made the most of this miserly offer by slicing the skin into such thin strips that it encircled an entire hill. Byrsa Hill* became the citadel for her new city of Carthage.

The grid squares I had visited so far encircled a large area after nine months of exploring. Riding mile after mile gave me a better overview of the topography of the landscape than I had grasped before from visiting individual fragments. I realised how close together some places were, and appreciated building up a better idea of the patterns of my map. And I became more aware of how all the rivers and roads and hills linked the land together.

How much do you know about your local area? What is the oldest woodland on your map? From which hills does the rainfall you drink come, and what watershed do you live in? Which rivers do your fertilisers and sewage leach into? Where does your rubbish go? What flower blooms first in spring? What does dawn smell like in the woods? The maps we inhabit, and the ways we live on them, are small but

**From the Greek word for ox hide,* βύρσα.

interconnected representations of the world at large.

I left behind a town's terraced streets and rode west towards the marshes, zipping down a canal towpath on my way to the very first square I'd visited, that inauspicious beginning when I drove my car into a ditch.

From that far corner of my map, I turned inland once more, swooping north down narrow lanes lined with full summer hedgerows. Short but sharp hills carried me through farmland and villages, bringing me to the high point of the area with views in every direction.

The sun burned as I hurtled down the map's longest descent, offering rueful grins of encouragement to a trickle of cyclists labouring in the opposite direction. Their trendy cycling clothes (£295 for some Lycra shorts, anyone?) and hipster beards suggested day-riders from the city, whose tall buildings I saw occasionally in the distance.

Among the many, many benefits of bikes over cars is the chance to enjoy the sense of smell when you're out on the road. It is an undervalued sense. Today's ride, for example, took me down a high street of frying chicken and past teenagers on a bench with the sweet reek of weed. There was the aroma of puddles and piss in a gloomy subway, and a muddy footpath along the edge of a golf course brought a waft of golfers' deodorant into the cloying scent of the sweet chestnut wood. As the afternoon warmed, I smelt the softening of tarmac roads, a charcoal barbecue at a bowls club, a car's faux-pine air freshener at a traffic light, the perfume of lavender fields, the fumes of exhausts, and my own sweaty stink.* I interpret maps visually, but this project would have been so different if I was focusing on the smells, sounds or tastes of a map. I would be fascinated to learn how a blind person experienced the grid squares I've documented.

I spun down from the hills through an industrial area with miles of warehouses and roadside burger vans. It made a nice change to ride fast, to feel my legs pumping and lungs working. Arriving at another

**Dr Kate McLean makes sensory maps like this, working 'at the intersection of human-perceived smellscapes, cartography and the communication of "eye-invisible" sensed data'. She leads 'smellwalks' around cities and translates the information into maps using 'digital design, watercolour, animation, scent diffusion and sculpture'. Her map of Edinburgh is titled 'Smells of Auld Reekie on a Very Breezy Day', while the description for Manhattan's map includes, 'Everyday smells include "gym people", "pretentious coffee roast" and "truck".'*

corner of my map, I turned ninety degrees again, hungry now, and cycled through a village of enormous houses and gardens with enviable views. This was a world away from the homeless shelter I'd ridden past a couple of hours earlier, though they were only separated by about ten miles in a straight line. You could find all of society on this single map, so close together, so far apart.

I slurped down a squashed banana as I rode, then paused at a newsagent's to stuff my face with sugar, salt and fat – all the easy calories. One of my favourite things about being a cyclist is to slump on a warm pavement, grimy and weary, and shovel food down my throat as fast as I can chew. Traditionally, I plump for a Coke, cheese and onion crisps, and a pack of Eccles cakes (which I never eat at any other time). I refilled my bottles from an outside tap and hurried onwards, refuelled and loving life.

I whizzed through wheat fields and housing estates, along A roads and sleepy byways with grass growing down the middle of the lane. I rattled down a pilgrims' way dating back to the Stone Age. A kestrel hovered overhead and a colourful lavender field was filled with young folk snapping selfies.

At one point, I seized the initiative from Google, convinced I knew a better route through the woods. Years of being on the receiving end of my own hapless shortcuts meant I was not at all surprised to end up hauling my bike through long grass, carrying it up muddy slopes and wriggling through dense thickets until eventually being faced with an

impregnable wall of nettles. I had done a poor job of combinatorial optimisation on my route! Too stubborn to retreat, I grabbed a stick and began slashing a way through the nettles. I got stung so much that I would lie awake for hours that night as burning waves washed up and down my legs and arms. It wasn't an entirely unpleasant pain, but it stopped me from sleeping.

Despite my 'shortcut', I reached the affluent, peaceful third corner of my map and hoped I still had time to make it all the way across to the final corner – an area of free school meals, brownfield development and busy roads. I set off at top speed, but had to concede defeat. The football would start before I made it home, and there was no way England could cope without me yelling advice, encouragement and insults at the TV. I left the remaining grid squares for another day and raced towards my sofa, singing loudly as I rode, 'I'm going home, I'm going home, I'm going, I am going home!'

The afternoon's miles had burnt, apparently, 2,119 calories, which equates to nine pints of lager and a packet of crisps. But I settled for a solitary, celebratory beer in a pub garden close to home. I was so enthused to have spent a summer afternoon zipping along under my own power that I didn't need more than that to cheer my soul.

And besides, England needed me. My timing was perfect. I hosed down my bike, jumped in the shower, and landed on the sofa just in time for the national anthems. Cheers!

BUTTERFLIES

'Will wonder become extinct in me?'

Henry David Thoreau

I removed my bike helmet and wiped my sweaty face. It was hot. I was at a memorial to a pilot shot down by German Messerschmitts in the skies overhead during the Second World War. Appropriately, the fields around were filled with poppies. Scattered at the base of the memorial was the rubbish from a KFC takeaway. The ten-piece Wicked Variety bucket contained 4,790 calories, the large fries had 1,440 and there were 750 more in the large Pepsi. I hoped it had been shared around, for that is a spectacular 6,980 calories, enough to fuel one eater through an impressive 69.8-mile run. Although given that they had been too lazy to put their rubbish in a bin, I doubted these calories were being used for long-distance running.

A cockerel crowed from behind a nearby hedge, jubilant not to have been fried. I rarely heard cockerels around here, but the sound reminded me of travels in other countries, of pre-dawn wake ups in the Philippines and the potholed roads of rural Nicaragua.

It was a hot, bleached-out morning, so I appreciated the brief burst of shade as I cycled under the motorway. Two workmen in helmets and one in a turban were wielding noisy power tools to add to the din. The smell of fried breakfasts wafted from a garden centre, while the waste ground alongside it was a riot of yellow, red, and purple wildflowers, popular with cabbage white butterflies. Farther down the road, I passed arable fields, grass meadows, a water-treatment works and an old mill converted into industrial units. There was such a variety of land usage in each of my busy grid squares.

A narrow footpath led between spiked fences guarding a fishing lake, where members pay £1,000 a year to catch 100lb catfish and 60lb carp. The path continued between an industrial unit and the motorway, lined with wilting cow parsley past its prime. Who, I wondered, would ever use this path? The first mushrooms were pushing through the dank earth, and bindweed vines spiralled up the fence. Bindweed spirals to the left, whereas honeysuckle loops right. Why? How? And what the heck prompted Flanders and Swann to write an entire song about it?

'The fragrant honeysuckle spirals clockwise to the sun,
And many other creepers do the same.
But some climb anti-clockwise, the bindweed does, for one,
Or Convolvulus, to give her proper name.'

I like bindweed's big, bold flowers, those white trumpets that bumblebees snuffle and forage inside. Yet gardeners loathe the plant, for it is notoriously hard to get rid of and its prolific growth can outcompete shrubs and even small trees. The roots creep underground for three metres and then send out runners that finger outwards until they touch something they can climb, and begin to lay down a new root system. And if that doesn't take over an area, the seeds can lie dormant in the soil for up to thirty years, biding their time and waiting to bind.

The path ran on, past the stink of a bonfire burning plastic, past Portakabins, hard standings, metal sheds and flatbed trucks. It linked nowhere with nowhere in the ugliest of ways. Then it spiralled left, like bindweed, up a ramp onto a footbridge across the noisy, smelly motorway. On the far side I found a scrapyard and mechanics' garage whose office was two shipping containers stacked on top of each other. A flapping tarpaulin kept off some of the sun's heat and provided a little shade for the enormous flat-screen TV in the upper container.

The office radio was playing 'The Living Years' (aptly enough by Mike and the Mechanics), a song that gave me an instant flashback to the classroom I was in when I first heard it. The power of songs to evoke memories is extraordinary, similar to the impact that smells have. For me, songs are often connected with travelling, and I had toyed with listening to music while exploring the grid squares this year. But, although I like the way music reminds me of places, I've always tended to use it on expeditions as a way of escaping mentally from hard times. Exploring this map was not about escaping, and I always tried to remain present and alert. So this meant no music and trying not to allow my phone to distract me with anything except the map and the informative apps I used on each square.

I passed a horse field dotted with yellow ragwort, a poisonous wildflower. Its bitter taste is off-putting so horses tend not to eat it, but ragwort is harmful if it gets incorporated into hay, where it loses its bitterness but not its potentially lethal toxicity. The government can fine people who don't control injurious weeds. On the other hand, it is a native wild plant that provides food for bees, insects and birds. Once again, balancing nature's needs with how we want to lead our lives is a difficult conundrum.

There have been so many issues on my map that don't have a simple answer. How to balance development with habitat loss, access versus privacy, farming tradition versus sustainable food, efficient ways to use the land, and so on. Such-and-such has irritated me, but I know you worry about something else. This matters to me, but that issue is more pressing to you if you're younger/older/richer/poorer than me. This is my world view, but does it come at the expense of yours? I've spent twenty years promoting adventure, but what is the best use of my next two decades? *There are more questions than answers*, sang Johnny Nash, *and the more I find out the less I know.*

I certainly felt more aware now than at the start of the project, about the range of demands on this landscape that needed to be considered, and I was less certain about some of my assumptions. But I also felt clearer than ever that the environment needs immediate action on all fronts: everything, everywhere, all at once, by all of us. Britain has been world-beating in destroying its natural landscape, and our population is the most divorced from nature in Europe. The issues are all related, and so are their solutions.

Most people agree it's important for us to move towards a greener world, but many would prefer that we only make slight adjustments at any one time and plan things carefully. That is often sensible in life, but in this situation we no longer have time. Nature and the climate are in critical danger, and we are making it worse every day.

If your bathtub was overflowing, the first thing you'd do would be to turn off the taps. We need to turn off the taps that are streaming harmful greenhouse gases into the atmosphere and heating our planet towards irreversible tipping points where the Arctic ice will melt, the Amazon rainforest will fail, and no carefully planned future will ever bring them back.

But we are not even slowing the flow of the taps: we are still cranking them up each year (bar a brief dip during Covid, which showed we *can* make major changes quickly if the desire is there to do so).

If your house was on fire, you wouldn't make a careful long-term plan about its restoration and repair. You'd put out the flames and then repair the damage. We need to put out the destructive fires of habitat loss and extinction right now, while also – of course – making thoughtful plans for the future.

Taking urgent and significant action as individuals and as a society is vital, both practically and morally. The way we vote, shop, eat, travel and live all has an effect. But moving to a greener world need not be a sacrifice; it can make our lives better. As well as holding our government to account, there is plenty we can each do that is clear, simple and impactful (see page 365). Best of all, many of these actions also generate joy, purpose, and bring us closer to nature again.

Summoning my nerve, I rode at top speed down The Nation's Worst Footpath®. Nettles jammed the path that was barely a metre wide and wedged between two high fences. It was like running the gauntlet on some comedy Japanese game show.* I escaped from the pain game of Nettle Hell® into a peaceful churchyard where people have worshipped since the 11th century, right back to when the Domesday Book detailed the taxable resources of all the boroughs in England. It recorded that

**In the 1990s,* 'Downtown no Gaki no Tsukai ya Arahende' *featured highbrow pain games such as 'Penis Machine', where contestants had to recite tricky tongue twisters while being punched in the nether regions.*

this village had sixteen villagers, ten smallholders and one slave, and owed taxes of £14 5s to the local lord.

Discarded Budweiser bottles were lined in a neat row on the church wall, perhaps by someone thinking that tidy littering was better than untidy. A gravestone amid the long grass was inscribed with an old Gaelic blessing:

'Deep peace of the running wave to you.
Deep peace of the flowing air to you.
Deep peace of the quiet earth to you.
Deep peace of the shining stars to you.
Deep peace of the gentle night to you.'

The church doors had been flung open to welcome in the summer breeze and any passing supplicants. I heard an elderly woman inside saying, 'I know people will think I'm odd, that poppies are only for Remembrance Day in November. But on this day thousands of young men went over the top in the Somme, so I'm putting out poppies.'*

A venerable yew grew in the churchyard, now bent with age, its trunk split and hollow. Beneath it was a gravestone with a carved inscription about the woman 'who caused this yew tree to be planted at the foot of the Grave of her Grandmother in the year 1742'. It must be quite unusual to know the precise age of an old tree, and I liked the idea of planting a tree when you bury a loved one.

Continuing onwards, I pushed my bike along a path through hay meadows parallel to the motorway. They were lush with flowering clover, though I failed to find a four-leafed clover, a symbol of luck for early Christians who associated it with the holy cross. More prosaically, but perhaps more helpfully, clover provides fodder and silage, as well as enriching the soil with its nitrogen-fixing abilities that can help lessen our fertiliser dependency.

Ancient people fed clover to warhorses, though if cows eat too much they can die from bloated stomachs. Pliny recommended its medicinal powers for poisonous bites, and the Chinese prescription of using clover to treat whooping cough migrated all the way to Europe, before crossing the Atlantic to North America with Dutch immigrants.

**The first day of the Battle of the Somme, 1 July 1916, was Britain's bloodiest ever. For the winning of just three square miles of land, fewer than eight of my grid squares, we suffered 57,470 casualties and 19,240 fatalities.*

Clouds of meadow brown butterflies busied themselves among the clover and purple knapweed flowers. I hadn't seen so many all year. Butterflies are often colourful, although they seem even snazzier to their own eyes. While our vision is limited to three primary colours, butterflies not only see UV light but are also tetrachromatic, meaning their eyes have four types of cone cells (compared to our three). They can therefore distinguish many colours that are imperceptible to us, a fact which blesses the butterflies' world with rainbows of colours that our dull and limited brains cannot even imagine.

I was glad to have come to this square on a summer's day, for the stream through the meadow looked very appealing. Everything looked very mellow, as no doubt appreciated by whoever had spray-painted 'Ganja is life' on the small wooden footbridge. I tried to imagine the grid square in an opposite season, with the fields bare and the stream in spate. What I find on each outing is true only on that day: visit again in winter, in pouring rain, or even in a different frame of mind, and my discoveries would be completely contrasting. But today, it was time for a dip.

Beyond the pastures, the stream ran wide and shallow beneath a colossal motorway bridge. Different murals have covered the walls each time I've passed this way. Last time there was a manic-looking Boris Johnson, ten-feet tall and depicted spraying an aerosol can labelled 'PANIC'. Boris was long gone, covered by a vast HELCH motif from

a graffiti artist who apparently annoyed the Queen by tagging the viaduct in Windsor, right in front of her castle.

The stream narrowed and then deepened before flowing round a bend into a wood. I stripped off and waded into the sandy shallows. Tiny fish flitted round my toes, and blue damselflies skimmed the surface. Green river weed waved in the clear, cool water and provided a soft cushion as I lay down and submerged my head. Opening my eyes underwater, I found myself in a dappled, blurred world. I listened to the deep peace of the running stream gurgling around me. The motorway felt far away down in the stream's womb.

Afterwards I sat on the riverbank in my shorts, drying in the sunshine as I drank the coffee I'd brewed on my stove, ready for when I got out of the water. I bloody love England in the summer sunshine. Right now, I felt no need for anything or anywhere beyond the flowing air and quiet earth of these few square kilometres of map.

FERRY

'There are things we will never see, unless we walk to them.
Walking is a mobile form of waiting.
What I take with me, what I leave behind, are of less importance
than what I discover along the way.'

Thomas A. Clark

To reach today's square, I needed to make a short crossing on a small ferry, which I knew would be fun but also added the *tiniest* fraction of hassle to proceedings, which is all I ever need to be tempted to procrastinate. That quibble aside, I always enjoy ferry crossings. The only thing that beats them are cable ferries across rivers, with a bonus point for those you have to hail by shouting, hoping that the ferryman hasn't gone home for lunch or closed for the season. Though these journeys are brief, they have the excitement of crossing a border, a boundary, to somewhere new.

Although today's river was only a few hundred metres wide, I wasn't brave enough to swim or canoe across it. The brown water swirled and boiled with eddies and undertows, and ships ploughed up and down. Even the ferry struggled, crossing the current in a wide, swerving arc.

As the ferry slowed down to dock, I looked back across the river at the landscapes I had been linking this year. I enjoyed seeing those connections from this fresh perspective, noting how this place joined onto that place. I wheeled my bike down a causeway of riveted girders, over tidal mud and shopping trolleys, then pedalled away from the ferry.

A dockyard took up much of the grid square, an in-between space that was simultaneously busy and deserted. 'No entry' signs abounded. There were no cyclists or pedestrians anywhere, but articulated lorries rushed around hauling shipping containers bound for every corner of the earth.

Even among the concrete and barbed wire, nature was fighting back. The verges grew wild, and brambles and bindweed scaled the security fences. Buddleia bloomed everywhere, an immigrant shrub that has settled successfully and is beloved by butterflies. Its long clusters of lilac flowers thrive in industrial parks, beside railways, and on waste ground everywhere.

Wild rocket, another naturalised plant, grew beside a vast lot of thousands of imported cars. Its yellow flowers look similar to ragwort, though the distinct peppery smell and flavour is even stronger than the stuff in supermarkets. I picked a few leaves to nibble as I rode around.

A flock of starlings was quite accustomed to the noise of passing lorries but was startled into flight by me and my bike. They whirled away over a scrap of waste ground where a forlorn tethered pony was munching its way through what little brown grass it could reach between the main road and the railway line.

Farther down the railway line was a row of scrap-metal dealers, protected behind high walls. The name of one business was roughly hand-sprayed along a corrugated fence next to a large sign saying 'Recycling the past for a better future'.

The residential town began on the other side of the tracks, opposite an enormous Amazon warehouse. I watched a workman watching another workman dig a hole, while outside a modern brick church an elderly man and his young grandson waited in shirt and tie for a funeral cortege to arrive. He looked at his watch then lit a cigarette, whiling away the time.

I pedalled around streets of unusual-looking, flat-roofed terrace houses and blocks of flats with peeling paint and the feeling of a marginalised community. One house had daubed a warning on its gate,

'NO SALESMAN BEWARE DOGS'.

Last week's award of The Nation's Worst Footpath® was usurped by an alley so blocked with nettles that it forced me to retreat. My flip-flops were admittedly not ideal for these expedition conditions, despite their proud claim to be one of the oldest forms of footwear: the trendy beachwear/terrible nettlewear feature on 6,000-year-old Egyptian murals, though back then they were made from reeds and papyrus, rather than today's ocean-floating plastic. Until the wise old Greeks began wearing flip-flops with the strap next to the big toe, as we do, the Romans and Mesopotamians wore them with the strap between the second and third toes. The modern popularity of flip-flops began after the Second World War when American soldiers returned home from Japan with pairs of *zōri* made from rice straw. Manufacturers copied them, launching a fashion trend that continues to this day.

Detouring around the impassable footpath, I rode towards the sewage works and a medieval artillery fort. A settlement of Traveller caravans and their dog kennels were tucked off the main road on a quiet lane. Their ponies grazed in the marshy fields, including a tiny one no higher than my knee. Clouds and blue sky reflected in the moat as I crossed the fields to the fort. Unfortunately, it was closed today. I was tempted to try to sneak over its chunky walls for a look around, but a fort is – by definition – tricky to get into.

I cycled along the riverbank, which was a high, concrete flood

defence, more like a sea wall than a river wall. I cast a suspicious eye at the pipework sticking out of the wall, knowing that the local water company had already dumped millions of tonnes of sewage into our rivers this year.

I stopped to talk to an elderly gentleman who was working carefully on a watercolour painting of the fort. I captured the same image in 1/640 of a second with my camera. He chuckled at the comparison. I wished him well and pedalled on, for it was time to catch the ferry back to my side of the river.

As we motored across the water, our little boat bounced over the wake of a container ship bound for the open ocean. I felt curious, as always, about who was on board, and envious of the places they were going. But I consoled myself with the thought that I too had been on a small journey to somewhere new today.

LAKES

'And above all, watch with glittering eyes the whole world around you because the greatest secrets are always hidden in the most unlikely places. Those who don't believe in magic will never find it.'

Roald Dahl

There was a humid, jungle feel to the day after heavy overnight rain. Plants shone, the ground steamed, a thrush sang a persistent tune that wouldn't have sounded out of place in the tropics, and pink rosebay willowherb flowers gave off their strong, sweet fragrance. The plant is known as fireweed in North America, and its scent always reminds me of it growing on blackened land following forest fires when I cycled through Canada. The dormant seeds make the most of the increased sunlight and decreased competition after fires, to bloom quickly before young trees return and outgrow them. During the London Blitz, willowherb was called bombweed as it flourished in the wreckage of buildings.

I was wearing shorts and a T-shirt today and not even carrying a raincoat in my bag. The weather had rarely been so clement this year and I had been excited to get on my bike this morning. Yet although the weather was kind, the overgrown footpaths continued to be anything but. This was another week of hacking through brambles, squeezing past nettles and swatting mosquitoes in damp undergrowth. All this slashing and whacking and stinging felt like a jungle expedition, albeit a gentle one accompanied by 4G phone signal and the sound of motorways. I'm content that this is about as ferocious as the British countryside ever gets.

A raindrop pearl jiggled free each time a carder bee alighted on a pink Himalayan balsam flower, also known as policeman's helmet, kiss-me-on-the-mountain and poor-man's orchid, as well as by its Linnaean name of *Impatiens glandulifera*. The 'impatience' comes not from its prolific growth – from seed to six feet in a single season – but from the explosive nature of the pods, which fire seeds for several metres to aid distribution.

Himalayan balsam has thrived here since its seeds arrived in Lancashire mill towns among sacks of cotton bolls from India. It spread along canals and riverbanks where, if left unchecked, it chokes the stream. This threatens biodiversity and erodes rivers, because when it dies back each winter there is little material to hold the riverbanks together.

Among these boisterous pink flowers were lacy clusters of creamy meadowsweet. John Gerard, a 16th-century herbalist, called it the 'Queene of the medowes' that 'delighteth the senses'. It was used as a strewing herb, strewn on floors to improve the fragrance of rooms in those pungent times.

An underused footpath ran into woods behind a row of houses, past a steaming compost heap of lawn cuttings that smelt of summer. When I was little, I used to enjoy poking through my dad's compost heap in search of frogs, and the smell brought back those memories. I carried on down to the lakeshore but was disappointed by signs warning of private property and forbidding me to swim. Normally, I would have heeded the mantra that it is easier to seek forgiveness than permission and just jumped in, but there were several fishermen sitting with rods and cool boxes, and I didn't want to disturb their day. It requires a bit of give and take if everyone is to enjoy the outdoors.

So I sat on a bench by the lake to eat my packed lunch and watch a pair of grey squirrels foraging. Then two rats emerged from the undergrowth and joined the squirrels. Although the animals were of similar size and appearance, the rats repulsed me in a way that the squirrels didn't and wild animals rarely do (exceptions: snakes, massive spiders, and any time I need to rescue a flapping little bird from inside the house). I have never had an unpleasant experience with rats, so this must be a learnt prejudice I have acquired.

The brown rat thrives wherever humans live, cheerily munching anything from insects and fruits to the contents of your bin and last night's ill-advised kebab. They dig burrows, live in loose colonies, and are prevalent around urban areas, where a female can rear a litter of up to twelve young, five times a year. There are twice as many rats as people in Britain, with more than 20 million in London alone. So you're never far away from the nearest rat, though possibly not the six-foot figure of urban legend.

I swapped the peaceful lakeshore for a path alongside the intimidating roar of a motorway slip road. Only a metal crash barrier separated me from the traffic. The path led towards the local allotments, which were overflowing with produce. Allotments are in high demand these days, with long waiting lists in many areas. Gardeners bustled around their patches, growing healthy food, keeping active, saving money and doing their bit for nature and the environment. Raspberries and beans ran up canes, onions swelled from the soil, and courgettes bloomed everywhere.

A motorway hotel's extractor fan was pumping out a mouthwatering bacon smell. Round the back, beyond the overflowing bins and the temptation of fry-ups, I found a footpath into a wood. Maple trees grew close together, and ivy wrapped around their trunks and spread over the ground. It was cool and dark under the leaf canopy.

A mound of pale wood dust caught my eye, covering everything on the base of a dead tree that was being devoured from the inside by an intrepid troupe of beetles. I couldn't spot any to identify what they were, but perhaps they were one of the many wood borers with splendid names such as the death watch beetle, ambrosia beetle, fan-bearing wood borer, or the Asian long-horned beetle.

Spiders had set up shop on the outside of the decaying tree, wrapping the trunk in silk webs. These, in turn, were dusted with frass (the technical term for wood dust), lending the tree a cheap Halloween effect.

Less spooky were the tiny flowers of enchanter's nightshade that thrived in this shaded woodland. The plant 'groweth in obscure and darke places, about dung-hills, and untoiled grounds, by path-waies and such like', wrote John Gerard in his *Generall Historie of Plantes* four centuries ago. Its Latin genus name, *Circaea*, comes from the woodland enchantress Circe, who cast a spell over Ulysses's crew and turned them all into pigs. I had never shown much interest in flowers before this year, but the names of wildflowers are delightful throwbacks to generations of people being far more connected to the natural world than we are today.

I then pootled up and down the residential streets of a 1960s development for a while. A pensioner was on his knees with a dustpan, brush and bottle of weedkiller, diligently removing every scrap of organic matter from his driveway (and, incidentally, from the skies above it, for the £5 billion global weedkiller market also reduces the abundance of birds). His lawn was an immaculate rectangle of fake plastic. This was a gentleman who liked order in his garden.

I confess I failed to find much of note in those orderly suburban streets, so I pedalled on to a park on the far side of the lake. The gravel path around the perimeter was busy with elderly strollers, professional dog walkers, one or two cyclists, a very sweaty jogger, some parents with prams, and even a scuba diver who emerged from the murky depths of

the flooded gravel pit – 8.3 metres deep, as he told me enthusiastically, dripping in his aqualung. Scuba diving is one of my favourite activities, though it had never occurred to me to look for it on my inland map.

I'm not really a fan of organised countryside. Perhaps it is the rules, the tidiness and order, or just being around lots of other people. I'm spoilt, I know, but I preferred the rougher, emptier corners of my map. But any green or blue outdoor space is important and beneficial, of course. And I was glad to see lots of people enjoying the fresh air and the water. The rewilding and transformation of this old industrial site had certainly benefitted the community, and the office workers who'd escaped here for their lunch hour were definitely enjoying the bonus fun of watching a soggy man in wetsuit and flippers waddle towards the carpark.

LUGHNASADH

AUGUST

VIEWPOINTS

'Find the good. It's all around you. Find it, showcase it and you'll start believing in it. And so will most of the people who come into contact with you.'

Jesse Owens

I sat down on an overgrown, underused bench outside a derelict timber-framed pub to squeeze out my socks. The men in hi-vis jackets from the water board had warned of a deep flood on the road, but I thought, 'Come on, lads, how deep can it be?' and pedalled on.

'Pretty deep,' was the answer.

Now I had wet shoes and socks for squelching around today's grid square. Well done, me!

Unique on my map, but very welcome, was a long strip of grass beside the road. It would not have been of much interest except that it was marked on the map as 'land available for access on foot'. Beyond the slender threads of footpaths and our declining municipal parks, this was a rare example of the 8 percent of England that is open-access land for anyone to roam freely.

The other side of the road had a line of houses whose names were a nod to former uses, such as The Old Post Office and The Old School

House. An elderly lady wearing a trilby and a nightie was smoking outside her front door. When she spotted me looking at her, I felt embarrassed, but she just grinned.

The houses backed onto a field dotted with red poppies among a crop of pale-blue linseed, or flax. Evidence from a cave in Georgia suggests that humans may have spun and dyed flax to make clothes as long as 30,000 years ago. Cotton production overtook flax in the 19th century, despite flax fibres being twice as strong. Today, flax is grown for its seeds and for linseed oil, whose smell takes me back to cricket bats, high hopes, nervous anticipation and disappointing dismissals.

I followed a path across the gorgeous blue field beneath singing and soaring skylarks. On the other side of a hedge, the change was dramatic. There were no more wavy blue flowers, nothing at all for hundreds of metres but the serried ranks of millions of wheat stalks, standing tall as they ripened towards harvest. I say 'standing tall', but cultivated wheat has become shorter and shorter. The late Norman Borlaug, an American agronomist, won the 1970 Nobel Peace Prize for his pioneering efforts in developing it. Semi-dwarf wheat, which produced more grain and fell over less often, was a contributing factor to the green revolution that doubled global yields in the 1960s and 1970s.

For millennia, crop yields were limited, as they depend on the levels of nitrogen in the soil. Traditionally, farmers used nitrogen-fixing

crops such as clover or soybeans to boost production, or spread seaweed, manure or guano (bird or bat poo, imported at great cost). But other than that, the only way to radically increase nitrogen levels before scientists began transforming what was possible would have been to stand in a thunderstorm with a long metal pole and hope to get your field zapped by lightning.

In 1909, German chemists Fritz Haber and Carl Bosch devised a method of converting some of the 4,000 trillion tonnes of atmospheric nitrogen into ammonia by reacting it with hydrogen at high temperatures and pressures. Today we generate 100 million tonnes of synthetic fertiliser by the Haber-Bosch process each year. Haber's discovery made him a hero and a Nobel Prize winner. It has been estimated that his breakthrough keeps 40 percent of humanity alive. Half the nitrogen in your body comes from it. Some say it was the greatest scientific discovery of all time, averting war and ushering in the modern age.

It is a pity that Haber didn't stop there, soak up the applause, and take up golf. At the outset of the First World War, the patriotic German signed up to help the war department. His expertise was so useful in the manufacture of explosives that his efforts may even have extended the war by years.

But things got even worse. Oh Fritz, why didn't you give golf a go? For Haber started playing around in his lab, added some chlorine to his ammonia, and duly invented poison gas warfare, and, indirectly, the development of Zyklon B, later used in Hitler's gas chambers. His horrified wife, chemist Clara Helene Immerwahr, shot herself dead in their garden.

Haber's friend Albert Einstein famously opposed the Great War, but in the Second World War he changed tack. He feared nuclear weapons but urged the American government to research the technology that his papers on quantum physics had inadvertently made possible, wishing Washington to develop nuclear weapons before Germany. In other words, good ideas can have bad unintended consequences. And that has certainly been the case with industrial farming since traditional techniques gave way to the green revolution's lashings of fertiliser, pesticide run-off, greenhouse gas emissions and continent-sized land clearances.

Einstein once said, 'We cannot solve our problems with the same thinking we used when we created them.'

And he also urged us to 'look deep into nature, and then you will understand everything better.'

Beyond the wheat fields I found an industrial yard filled with heavy equipment and cranes for hire. A scrap-recycling business had piled up pyramids of old tractor tyres and rusting heaps of complicated-looking machine parts. Vehicles in various stages of decomposition lay in the trees beyond the yard. The cab of a 1970s milk delivery truck had rusted and collapsed, leaving the steering wheel sticking up in the air. Its smashed headlights stared into tangles of brambles and the door hung limply from one hinge.

A swallow swooped low right in front of me, trailing more weather folklore in its wake. 'When the swallows fly high, the weather will be dry.'

There is a measure of truth in that, for swallows do hunt at higher altitudes on dry days when rising warm air sweeps up flying insects. Their prey seek shelter during cold or wet weather, so the birds need to swoop down to catch them.

'I love the swallows,' said the farmer I met at the end of the lane, looking up with me. 'I love all of nature, mind you. When I was a student in the city, I thought I'd go mad. It's an illness, you know. I needed to get out, somewhere green.'

This was the first common ground we had found in about ten minutes of polite but heated conversation. We both treasured swallows and nature and this landscape, but our argument stood between us understanding each other.

It had begun when I lifted my bike to cross a stile and walk up a footpath across some grazing fields.

'What the hell do you think you're doing?' called an angry voice.

I turned to see an elderly farmer stomping my way in wellies and a flat cap.

'I'm taking the path up to the wood,' I replied.

'Well, I'd really rather you didn't,' the farmer went on. 'I've got cows on this farm. The grass will get trampled, and anyway, the grass is wet. You'll get soaked.'

'I don't mind damp grass,' I replied in a conciliatory voice. 'I won't bother your cows or leave the path. But it's a public right of way. I'm allowed to be here.'

'I know it's a footpath, but it is a bloody nuisance and it would cause me deep distress if you use it...'

And off he went, on a long and well-rehearsed rant about the scourge of people coming onto his land, dogs off their leads, trail bikes scaring the cattle, and the 'bloody Ramblers Association going on about access rights'.

Arguing with an angry old farmer wasn't going to change his point of view, but the lack of access to the countryside I live on had been one of the repeated frustrations of my year on this map, and I told him so.

That was when he told me about the swallows and how important nature was to him. I replied that the 99 percent of us who don't own the country also need green spaces. He agreed, but then dived back into another tirade against people using the footpaths across his farm.

During our futile but slightly cathartic conversation (for me at least, though I suspect not for him), we established that we had a mutual friend – another farmer. At this point, he became extremely apologetic for getting angry with me.

'I'm afraid I haven't treated you in a very Christian manner, have I? Would you like to come and see the cattle, or have a cup of tea?'

The change of gear once we discovered our common ground was fascinating. I was no longer an undesirable 'other' but somehow now more legitimate. I thanked him for the invitation but declined. I'd be on my way. Not the way I wanted to go, a way I was entitled to go, but a way that wouldn't upset an old gentleman, and that felt like the decent thing to do.

I looked up at the acrobatic swallows as I rode towards home, free to soar and swoop wherever they wished, and wondered how much of the resistance to increasing land access is just a mistrust of people who are different.

POLYTUNNELS

'I wish to speak a word for Nature, for absolute freedom and wildness, as contrasted with a freedom and culture merely civil – to regard man as an inhabitant, or a part and parcel of Nature, rather than a member of society.'

Henry David Thoreau

I filled my bottles with ice before heading out this morning. It was the hottest day of the year, and Britain was parched by an unusually severe drought. As I got ready, I heard on the radio that twenty centimetres of rain had fallen in an hour in Germany, causing floods that killed almost 200 people.

The last of the morning dew felt cool on my toes as I cycled down a grassy path in my flip-flops. In the crisp, brown fields, the harvest seemed to be ripening before my very eyes. A silence hung over the day, which reminded me of Spain. A distant voice carried from across the fields. I was roasting. And it was still early. I envied a buzzard whose feathers ruffled in a breeze as it perched on a pylon by the railway.

These were the dog days of summer, hot and heavy, and so-called because the star Sirius, part of the Canis Major (greater dog) constellation, rises alongside the sun at this time of year. Greeks and Romans believed that the combined heat of these two stars accounted for the season's high temperature.

I clung to shade wherever possible. The roadside wildflowers were past their best now, wilting, hangdog, dusty and tired. In the distance I saw the first combine harvester of the season, as big as a tank and churning up and down the hillside. The farmers would be relieved when they completed the toil of gathering the harvest in this heat.

Harvest festivals have been celebrated since ancient times, including Lammas, or 'loaf day', which is one of the oldest connections between the Church and agriculture. It is a ritual celebration of the onset of harvest, marked with a loaf of bread made from the fresh crop.

This week marked Lughnasadh, the traditional Gaelic celebration of the midway point between the solstice and the equinox. High summer, and the start of the harvest season. A time to enjoy the warm, easy days, but also to prepare for the changing seasons to come. Historically, people celebrated with athletics competitions, feasts and matchmaking. It was started, so the myth goes, by the Irish god Lugh as a funeral feast for his mother, Tailtiu, who died from exhaustion after clearing Ireland to make the land suitable for agriculture. The first solemn cut of harvested corn was therefore always offered to Lugh by burying it at the highest point of the land.

The major feature I encountered on today's hot grid square was polytunnels, hundreds of metres of plastic tunnels filled with raspberry plants. The tunnels were not attractive in the way that traditional farmland appears to be at first glance, but they are efficient and play a significant part in the farming future we need to move towards, using less land to produce more food in less harmful ways.*

I enjoyed discovering that the footpath actually ran straight down the middle of one of the long tunnels. It was ferociously hot in there, so I sustained myself by foraging (aka stealing) a few raspberries. Although I felt a little guilty, I was surprised to see how much fruit was already rotting on the plants. In Britain, we waste 3.6 million tonnes of food annually before it even leaves the farms, and 9.5 million tonnes in total,

**Polytunnels generally use less pesticide, fertiliser and water than standard farming. They can grow crops all year round at higher yields, and are an excellent opportunity for using renewable energy such as geothermal heating or heat exchanges. Yet, as always, there are problems too. Dutch farms, for example, who are polytunnel pioneers, emit above-average ammonia per hectare.*

worth more than £19 billion. Supermarkets impose their obsession (or, more pertinently, *our* obsession) with the availability and appearance of produce on farmers, who are therefore forced to grow a surplus and can't do much with the leftovers.

Before the polytunnels and the conversion of nearby farm buildings into 'exclusive residential developments for the high demands of modern living', this square had been covered in cherry orchards. A few isolated cherry trees were still dotted around, ghostly reminders of earlier land use. The ebb and flow of the squares I have been treading this year has been fascinating. I passed a collapsing barn that was so enmeshed in ivy that its collapse and disintegration seemed to be happening in front of me. My map is fluid, flexible and fixable.

It is not always a one-way path towards the end of wildness. That industrial barn is returning to nature. This hay field will not stay this way for ever. The road through the woods may once again become coppice and heath. What has been will be again. The new railway line watched by the buzzard will one day be old, one day be obsolete, one day be strangled in brambles, and one day be here no longer. One day, this town may become a wood, or lie beneath the sea. Generations come and generations go. Humans will not be here in perpetuity. And one distant day, earth will return to being a dry, lifeless rock like it used to be, spinning silently until the sun runs out of hydrogen, expands into a red giant and swallows our planet. Far from being depressed by this, I found that focusing on the pinprick of time which I was so privileged to enjoy in this summer grid square was both reassuring and uplifting.

I followed a narrow footpath along the edge of a field of wheat that scratched my legs, then round a softer field of linseed into a cornfield. I hadn't seen maize growing on my map before. The tall plants looked tropical, their waxy leaves a darker shade of green than the other crops around here.

Corn is one of the world's most important crops, though we eat less than 10 percent of the billion tonnes grown annually it in its direct form. We eat a little popcorn and corn on the cob, but most of it is turned into the flour and syrup pervasive in processed foods, fed to livestock, or converted into ethanol and plastic.

Corn first became a staple foodstuff in the Americas, domesticated as long as 9,000 years ago in Mexico from a wild grass called *teosinte*. It is easy to grow and to store, and is very nutritious. Hundreds of generations of farmers selected the biggest and juiciest kernels, leading over time to modern corn with its gigantic, tightly packed cobs. The plant has been modified so much that it would no longer survive in the wild and we have become dependent upon each other. Although Columbus brought corn back to Europe 500 years ago, it only recently became a significant crop in Britain, where, as well as being used to fatten cattle, it provides cover for pheasants.

At harvest, a farmer may leave uncut sections of maize round the perimeter of fields to act as cover for pheasants reared for the local shoot. Fifty-seven million game birds are released and shot each year for sport in Britain (using 4,000 tonnes of lead), far more than in any other country. That equates to more than twice the biomass of all our wild birds, which inevitably makes it harder for them to compete for resources, although at least the shoots' land management does also benefit the wildlife.

I saw a blob of blue on my map and headed to investigate, hoping on this hot day (with extreme naivety) that I might find a clear, deep pool to swim in. No such luck. It was an especially disappointing pond, more swamp than water, and ringed with officious signs. Some moron had hurled the lone lifebuoy out onto the mud. Traffic hurtled along a busy road a few metres away. One sign informed me that this pond was, alas, no longer even a pond. It was now a PCD, a pollution control device, which can be sealed off in case of an accident to prevent spillages seeping farther into the natural environment.

The lane I rode along from the pond was more attractive. It was lined with countless flowers, mostly yellow oxtongue and white wild carrot, also known as Queen Anne's lace, with a single, tiny red flower at the centre of each head of white florets. I listened to the whirring click of grasshoppers among the wildflowers. There are about 25,000 species of Orthoptera, of which just eleven are native to Britain, although around thirty now live here.

Male grasshoppers perched on blades of grass, enjoying the summer sun, while, on the ground below, ants scrabbled to put away stores for winter. The grasshoppers were performing their mating calls, competing with each other through increasingly complex songs. Females then express their preference by mimicking the song they prefer. It is like a battle of the bands plus a karaoke contest.

I cupped my hands and caught one, then peered at it carefully through a gap in my fingers to stop the grasshopper leaping clear. The insects' powerful back legs have rows of bristles, like a comb. Rubbing them against hardened veins on their wing casings produces the cheerful chirring sound that they have been belting out since the days of the Permian moss forests, 230 million years ago.

Much of today's square was engulfed by a no-man's-land beside the railway tracks and the dual carriageway; off-limits to me, but useful corridors for nature. I crossed the dual carriageway via a bridge, then angled left up a hard-to-spot footpath onto an area of hay meadow dotted with blue wildflowers and butterflies.

I was surprised how many 'feels' this one square kilometre had. This grassland felt very different from the degraded pond and the fenced-off main road, and different again from the polytunnel farm. It was an unexpectedly rural expanse of land up here and I made a mental note to return for a run one day. Finding new running routes had been a nice spin-off benefit of this project.

As I followed a thin, straight footpath through the grass towards two mighty pylons, I met a lady walking her dog. Encountering anyone using the footpaths on my map was rare, so I always said hello. But today I was also moved to tell her how much I liked this open space I was surprised to have found.

'Yes, it's nice up here,' she replied. 'But it used to be a riding stable for the kids before they put in the railway line. It's a shame that it's gone.'

It was interesting to learn how the land had changed and to hear varying opinions about my map. I generally felt there were too many stables taking up too much space for the benefit of too few people. But I'm sure horse riders would say the same about golf courses, golfers about football pitches, footballers about shopping centres, and shoppers about woodland.

I always had to remind myself that my own preferences were not 'correct'. This map of 'mine' was also home to thousands of other folk. We all need homes, transport, food and employment, as well as space to exercise, headspace to think, and an aesthetic consideration for nature. Everyone has their say. Everyone except the land itself, I suppose.

However, more countries are now starting to give legal rights to ecosystems and nature. Ecuador and New Zealand have conferred personhood status upon rivers and mountains, increasing their rights not to be persecuted or violated, and the first court cases have begun in the US. It is not a new concept: Bogd Khan Uul mountain in Mongolia has been venerated and protected since the 13th century. What has been could be again.

Food for thought as I circled back towards home down the raspberry polytunnel footpath, perhaps just nibbling one or two more berries as I passed…

STREETS

'I am no scientist, but a poet and a walker with a background in theology and a penchant for quirky facts.'

Annie Dillard

I had waited for the rain showers to pass before heading out today, but I was forced to shelter from a fresh cloudburst beneath a bowed old horse chestnut tree. Sheets of water slid down the road and dampened my enthusiasm. I had, however, spotted the map symbol for a pub on today's grid square, and I had little to do later.

'Go for a look around the square, and after that you can go to the pub,' I bargained with myself.

It had been a warm and humid day between the heavy showers. Aside from traditional British grumbles, which we all enjoy, the weather had not actually been too bad recently compared with, say, the year 1816, when ash clouds from a volcanic eruption in Indonesia shrouded the world in an extended winter. Mount Tambora's blast was heard 1,600 miles away and plunged the 350 miles around the volcano into darkness for two days. It was the most powerful volcanic eruption in recorded human history.

Over the next year, a cloud of ash spread through the atmosphere, wreaking havoc with the weather for three years. The resulting potato famine in Ireland led to a terrible outbreak of typhus and mass emigration. North America's arable economy crashed, causing the panic of 1819 that pushed the country from being a commercial colony towards becoming an independent economy. In China, three consecutive harvests failed, prompting farmers to plant poppies in place of rice, with far-reaching and long-lasting global consequences.

But while Tambora's eruption caused widespread famine and disruption, the strange weather also influenced an output of poetic and musical works infused with gloomy genius and named for the Greek god of fire: Byron's *Prometheus*, Mary Shelley's *Frankenstein* or, *The Modern Prometheus*, and Schubert's first commission, the cantata *Prometheus*, composed to a poem of the same name by Goethe.

Volcanoes erupt now and then, and weather conditions also swing back and forth naturally, but sane people are in agreement that human behaviour is now causing climate breakdown far beyond natural variations. A clear and alarming demonstration of our extravagant and irresponsible way of life was the occurrence this week of 'Earth Overshoot Day'.

Earth Overshoot Day marks the date when humanity's annual demand for ecological resources and services exceeds what the planet can regenerate in that year. It means we've used up our sustainable biocapacity for the year. We deal with the deficit for the rest of the year by borrowing from the future and gobbling limited reserves of ecological resources more quickly than they can be replaced, if at all.

Qatar and Luxembourg's Overshoot Days for the year were back in February. Britain's was in May. The only reason the world's Overshoot Day as a whole is as late as August is because the poorest countries are still living within their means. They prop us up, while also bearing most of the burden and consequences of climate change.

Sustainable living dictates that you must meet the needs of the present without compromising the ability of future generations to meet their own needs. We are clearly failing to do that. How long would you tolerate the behaviour of a friend who guzzled voraciously, overspent in his own interests, then came to you each August asking you to bail him out for the rest of the year?

Once the rain passed, I left the shelter of the chestnut tree and

pedalled on to my grid square. Half of it was covered in sleepy suburbs whose residents took pride in keeping their gardens neat and their cars clean. One home had replica Roman statues on their white gravel drive. Another flew the English flag from a tall pole. A third had a gleaming vintage American Airstream caravan parked outside.

I began looking at the house names. Some settled just for the street number, but many had a name sign too. Some appeared to yearn for nature ('The Laurels', 'The Oaks', 'The Glade') or wilder landscapes ('Min y Coed', 'Langdale'). Others were whimsical ('Dewdropinn') or simply satisfied ('Finally').

Naming habits have evolved over centuries. When few people could read, shop merchants used to display symbols to show their trade – a civet cat for a perfumer, say. Two standing lions holding a pretzel and a sword meant a baker. A bull's head and two axes was a butcher. The oldest written tale in English, *Beowulf*, features a mead hall called 'Heorot', meaning White Hart, which is still a common pub name. Before 1200, houses were usually named after the owner, such as 'Ceolmund' or 'Wærman'. This shifted towards taking inspiration from heraldry, for instance 'Le Griffon', 'La Worm' or 'Le Dolfyn'. In the Enlightenment, house names leaned towards emotional states. Frederick the Great, for example, named his favourite palace '*Sans Souci*' or 'No Worries'.

I saw that a few houses had installed electric car-charging units.

But as most homes had two or more cars parked outside, it is clear that Britain needs massive infrastructure investment if we are to successfully ban the sale of new petrol and diesel cars by 2030. Doing so is part of the vital, swift transition we need to make towards electric vehicles (along with driving less), as transport currently accounts for around 20 percent of CO_2 emissions.

For this change to be possible, however, we will need far more vehicle charging units, until they become as ubiquitous as postboxes or, in their heyday, phone boxes, earlier emblems of technological and societal revolutions. Because Britain is small, it should be easier to create a nationwide infrastructure for electric cars than in large countries, but time is running out.

The average length of our journeys is just 8.5 miles, so most drivers only need charging top-ups at destinations such as supermarkets, especially those who don't have driveways or garages with their own charging facilities. Full charges for long journeys are needed much less frequently, but range anxiety is still one of the main issues slowing the switch to electric cars. It is generally more of a psychological issue than a genuine problem, as both driving range and charging point availability are increasing all the time.

But while there were 1.3 million public charging points around the world in 2020, and it is estimated there will be more than 16 million by 2030, both the number and the installation rate needs to increase significantly if the switch to electric vehicles is going to be successful.

I headed down the grid square's only footpath, a narrow alleyway leading out to the edge of the houses. I leant on a five-bar gate and surveyed a scruffy patch of waste ground blooming with wildflowers and busy with orange-tip butterflies. To my eye this all looked nicer than the swept patios and stripy lawns I'd just been riding around.

I wondered how butterflies managed to fly in strong breezes like today's. Scientists tested this by putting them on tiny leashes in wind tunnels, training them to fly towards flowers and filming their efforts. In headwinds, butterflies push forward by 'clapping' their wings behind their backs, twisting them to create additional lift through the whirlwinds that roll off their wings.

Headwinds are tough, but tailwinds can benefit butterflies, helping them to fly at 60mph, hundreds of metres above the ground on their

epic migrations, such as the painted lady crossing the Sahara Desert to Europe or the monarch butterfly's 3,000-mile flight across North America to hibernate in the forests of central Mexico.

I had been learning this year to consider where I lived as an interesting natural habitat in itself. While I have been lucky to visit many of the habitats you usually see on nature documentaries, wildlife also thrives in the suburban or urban environments where most of us spend our days. New housing developments, tidying up scruffy areas, and declining food supplies add extra stress to these environments. But I still regularly spotted carnivores like foxes, and saw fabulous birds such as herons mooching around in manky streams or sparrowhawks doing battle with pigeons. I could watch butterflies every day in the summer and see plants such as shepherd's purse, red dead nettle and creeping wood sorrel battling through cracks in the concrete. It is fantastic that millions of us live alongside such diversity.

I passed a playground of children enjoying their holiday freedom, then continued into farmland. It had been a challenging season for farmers, waiting for their wheat moisture levels to drop to the optimum percentage for cutting, only for more rain showers to set them back and get them looking anxiously at their harvesting contracts.

I searched unsuccessfully for the ancient boundary stone marked on my map. Instead, I found a small bench, obscured by a leaning sloe tree, and sat down to look out over the fields. This grid square consisted almost entirely of the streets I'd ridden around and these enormous fields of wheat. There was not much variety or scope to follow my nose.

I was pleased, therefore, to stumble upon some community allotments. I don't yet have the patience to be a gardener, but I think one day I will enjoy growing my own vegetables. Digging up potatoes and picking tomatoes is very satisfying. I enjoyed seeing how each person had used their allotted space, and wandered around admiring the rows of onions, cabbage, corn, pumpkins and sunflowers.

Allotments are measured in units of rods, also known as poles or perches, an obscure 5.5-yard measurement dating back to the 14th century. (*Tres pedes faciunt ulnam, quinque ulne et dimidia faciunt perticam*: three feet make a yard, five and a half yards make a perch, obviously). They were originally ten rods long, an area about the size of a tennis court. The ones today were much smaller, sub-divided to allow more people to have an allotment and to make them more manageable.

While allotments have been around since Anglo-Saxon times, the system used today dates from the 19th century when labourers were allotted some land to grow food. There was no Welfare State and the country was becoming more urbanised, so being more self-sufficient was important.

Not all the early allotment rules have survived the test of time. Sunday is now the most popular day for gardening, yet in 1846 allotment holders were 'expected to attend divine service on Sundays; and any occupier who digs potatoes or otherwise works on his land on Sunday shall immediately forfeit the same'.

Pubs have also relaxed their opening hours since those days, for better or for worse. I checked the time, shrugged to myself, left the allotment, and cycled to the pub.

DAYBREAK

'That night was the turning-point in the season. We had gone to bed in summer, and we awoke in autumn; for summer passes into autumn in some unimaginable point of time, like the turning of a leaf.'

Henry David Thoreau

It seemed to me, walking and cycling through this year on my map, that the seasons move in two ways: gradually, then suddenly. No change, no change, no change… and then one morning the new season is well on its way, overlapping the previous one in its eagerness to get going. I caught the first embryonic smells of autumn today, along with heavy dew and a noticeably later sunrise.

I always enjoy daybreak, though doing the school run means I'm rarely free to head out and play at such an hour. But I managed it this morning and immediately felt I was winning the day. It took an hour to ride across my map to the grid square, and I had time to enjoy the sun rising, the rabbits in the fields, and the foxes slinking home after a big night out.

Eventually the road narrowed, grass grew through the middle, and a sign said 'No Through Route'. I kept going. A woodpecker chattered in a shaded pocket of woodland. The road dwindled to a track. I kept

going. I often wish I lived somewhere wilder, but within my map, not too far away, I can find places like this that feel gently wild. The wild places are not as distant as I often fear. This year had made that abundantly clear to me. I just sometimes have to look closely to find them.

The track became too steep to pedal, or at least too steep if you were in a late summer, laid-back kind of morning mood. I pushed my bike the rest of the way, enjoying the cool shade as I entered a beech wood. Despite this wood being an SSSI, there were still drinks cans and half-burnt bags of rubbish flung around the undergrowth. Trees covered the steep slopes of much of today's grid square, with meadows on the plateaus and valleys. It was one of the most rustic and scenic corners of my map.

The path emerged from the woods on a ridge and crossed into farmland. A sign warned 'Bull in field', but today I had the place to myself, a picturesque patchwork of small-scale farming, with mixed crops of corn, grass and wheat divided by hedgerows.

It was still early but already warm and shaping up to be a glorious day. I swooshed down a footpath through a bumpy field of dewy grass with crickets bouncing out of the way of my wheels. A familiar but hard-to-place fragrance caught my attention and I squeezed the brakes to investigate. Then it came to me: pizza! I looked down and noticed the small pink flowers carpeting the ground. It was wild marjoram, the herb called oregano, which is so familiar in Mediterranean cooking. Marjoram grows prolifically in chalk grasslands during the summer

months, and riding over it was jolting the fragrance into the air.

I was snapped from my hillside botany by the nasty nip of a horsefly. Their bites hurt, but the flies are slow, and I gained a modicum of revenge by splatting it before it could escape. My new nature-loving zeal had its limits. Only female horseflies bite like this, for they need the protein from blood to develop their eggs. The bites hurt because the flies just saw callously through your skin with their serrated mandibles until they strike blood, and they don't administer the mild anaesthetic that accompanies a mosquito's more precise, surgical injections.

Carrying on down into the valley, I was intrigued to come across the site of a lost village at the end of a quiet track. The term 'lost village' generally refers to the 3,000 villages which were abandoned or emptied in Medieval times, many as a result of populations being wiped out by the Black Death. This was sobering to consider with the awful Covid years still fresh in my mind.

All that remained of this lost village was a tiny church that had changed little in nine centuries. It was a modest stone building with a straw-covered floor, a stone arch and an oak-framed roof. It was staggering to think it had already stood silent and empty for 150 years by the time Columbus sailed the ocean blue in search of popcorn. The nooks and crannies of my map carry the centuries very lightly.

Today, builders were hard at work outside the church, bringing some 21st-century vibes into the valley with their cement mixers and Magic FM. But they were modernising in an antiquated way, constructing a columbarium in the church grounds. A builder explained the plans to me in enthusiastic detail. A columbarium is a place for storing cremation urns, popular since Roman times. This new one was being constructed in the style of a Neolithic long barrow, albeit using 180 tonnes of concrete, sunk beneath the ground. Who knows? This may be the start of another thousand years of history for future generations of local families.

SWIMMING

'People say you have to travel to see the world. Sometimes I think that if you just stay in one place and keep your eyes open, you're going to see just about all that you can handle.'

Paul Auster

I cycled to today's grid square with *Test Match Special* playing in my headphones. Listening to the ebb and flow of a cricket match arcing towards its conclusion is one of my greatest pleasures. I turned it off reluctantly when I arrived so that I could concentrate on what I was exploring.

I began outside a working man's club with a fluttering Union Jack, then rode among Victorian terraces, streets of post-war pebbledash, and 1980s semis. A brick clock tower had been built in the town centre with the largesse of the local mill owner 150 years ago, and the mill's chimneys still smoked away in the distance. There was the usual array of shops and eateries: convenience stores, kebabs, fried chicken, Chinese, Indian, garage doors (that was a first), and a bookmaker. It was a typical old-fashioned town of struggling shops and pubs sliding into decline, plus a shiny new Domino's Pizza takeaway.

An elderly man laboured across the street with his shopping trolley.

A car slowed and waited an age for him to cross. 'That will be me one day, sliding into decline,' I thought to myself. 'Be grateful then for this moment. This moment is my life.'

The town was more or less as I'd expected when looking at my map before cycling here. What I hadn't anticipated was the new area of clapboard houses and commuter apartments built behind these tired streets. There were homes painted in cheery pastel shades, drainage culverts landscaped into trickling streams, and pots of geraniums. The gentrification even extended to an M&S store in perhaps the poshest petrol station on my map.

The clean uniformity of these new residential streets reminded me uncannily of suburban America. None of the building styles matched the traditional conventions of this region, but I was undecided whether that was a bad thing or good. I pondered whether or not this new town would merge with the old, if its injection of money and vigour would help the high street. Or would the vibe here overtake the traditional terraces down the road, with cappuccino to go replacing a cup of tea in the caff? Housing issues were yet another topic I had never considered before this year, but now found surprisingly interesting.

A flyer stapled to a noticeboard urged residents to take an interest in Hedgehog Awareness Week, declaring, 'We've lost a third of all our hedgehogs since the millennium. In an increasingly urbanised Britain, we choose to lose all that is complex and beautiful if we do not stand up for our wild animals and plants.'

On the edge of town, I found a scrapyard surrounded by caravans and trailer homes. I peeked into the photogenic and intriguing space but felt nervous about taking my camera out and exploring further. Instead, I headed up a path to a park, spotting on the way a mislaid badge lying on the ground, which read, 'Sorry, I need to lip-read. Please be patient.'

Should I now research the history of lip-reading ('The earliest record of deaf education was in 1504...') or the history of badges and brooches ('The earliest manufacture of brooches in Great Britain was during the period from 600 to 150 BC...')? Exploring these grid squares was exhausting: there was so much to see all the time.

Behind the town of two halves was a cluster of small lakes in flooded old chalk pits. Back in the day, lime taken from this square had been used to build the city's bridges and landmarks. My nose picked up the delicious scent of freshwater (take that, sharks!) and my hopes rose in anticipation of a refreshing swim. But high fences thwarted me, of course. It's the hope that kills you.

Signs warned of the insidious hazards of 'Deep Water' as I dreamed of the delights of deep water and the pleasures of wild swimming. Sealing off nature to protect us is ridiculous and counterproductive. Far better, surely, to teach people to be safe, to swim responsibly, to make sensible judgement calls, and then to open up the countryside to be enjoyed and valued.

If I lived in this town but couldn't swim or kayak or SUP here, I would be really frustrated. I suspect, however, that fencing off lakes and declaring nature off-limits is so ingrained as to be rarely questioned. And yet I knew that the local swimming pool has queues out of the door and rationed time slots, despite swimmers having to jostle in the crowded water and get changed in rooms that smell of wee. Open expanses of freshwater sound much more appealing.

I was pleased, then, to find that two fenceposts had been forced apart, allowing me to squeeze through and slip into the lake for an illicit swim. Nothing beats cool water on a hot day – the fragrance, the sunlight dancing off the surface, the rippling sounds, and the refreshing feel on your skin as you sink beneath the water and slough off the day's heat and frustrations. I like to open my eyes and dive down deep through the soft green light, feeling the temperature fall and the colours dim as I try to reach the bottom, and then pushing up for a triumphant gasp of air with a handful of gravel as proof of my journey into another world.

I towelled myself dry with my T-shirt, then left the lakes and the town and crossed a busy road towards a river and marsh. The road was lined with anonymous warehouses with thousands of plastic crates stacked against them. A dramatic missing-cat notice was sellotaped to a fence.

'On the night she went missing, at around 2.20am, our neighbour's CCTV caught someone acting suspiciously outside our home. Our worst fears are that this person may have had something to do with her disappearance.'

Unable to shed light on the feline mystery, I continued on my

way, ducking as I rode through a low underpass beneath the railway into an area of reedbeds, home to reed buntings and sedge warblers. Damselflies flitted in the sunshine. Pink and white marsh mallows grew along the verge, a plant fond of damp areas. As well as an ingredient for tooth-rotting deliciousness, the marsh mallow root also has medicinal properties. Early Arab doctors applied it as a poultice for inflammations, and it has been used as a laxative and to treat sore throats.

'Is tasteless food eaten without salt, or is there flavour in the sap of the mallow?' asked the Book of Job in the Old Testament, for mallow isn't very tasty until you mix it with loads of sugar. During the siege of Jerusalem in 1947/48, starving residents were forced to forage for mallow, known as *khubeza*. It was considered a weed but is high in vitamins and iron, so the local radio station broadcast ways of cooking it.

Four thousand years ago, the Egyptians boiled the mallow root and added honey to the mix, creating a delicacy that was reserved for gods or royalty. The French began making modern marshmallows in the 19th century, mixing the mallow's root sap with egg whites and corn syrup to create a fluffy mixture that could be heated then poured into moulds to set before being toasted on campfires by kids of all ages.

The footpath followed the route of an old industrial railway line, with hunks of concrete from demolished factories still dotted around among the scrub and bushes. I sampled a few blackberries from a hedgerow. They were getting sweeter, but still weren't quite ready. Autumn was on her way, but not yet here. I reached a bird hide, a simple wooden wall of planks with two horizontal gaps, one at adult eye height and one for children. I saw nothing more exciting than a few pigeons, and demanded a refund from the information sign that boasted of kingfishers, teal, egrets, water rail and gadwalls.

I carried on through tall reeds until I reached the river, which ran broad and slow and glimmered grey and gold in the sunshine, like lava. The tide was out, and the exposed banks were slabs of mud dotted with discarded tyres and an upturned car seat. New houses lined the far riverbank, with banners declaring them to be 'a unique and ambitious development, offering all of its residents a wealth of inspiring lifestyle opportunities'.

The sky was big, blue and empty here, the factory chimneys smoked away in the distance, and the only sound was the breeze rustling the reeds. There was nobody around, except for a father and son, who did a bad job of hiding catapults behind their backs as they walked past me. I wasn't sure what they were taking potshots at, but they clearly didn't want me to ask.

I liked it out here on this small marsh, tucked into a tight curve of the river that marked the boundary of my map, more or less. The other bank was unknown ground, off-limits in this project, and therefore looked extra tempting to explore. As fascinating as I was finding my map, the call of the horizon always lurked within me. So, too, did my desire to find out how the cricket was going. I tuned in to the match as I cycled home, eagerly hoping that England's contingent of Yorkshiremen had saved the day, as usual.

SEPTEMBER

ACCESS

'The world is full of obvious things which nobody by any chance ever observes.'

Sherlock Holmes

My hopes were high. It was a perfect sunny day and the grid square looked enticing on paper. It was mostly woodland, with some contour lines, a small lake and the site of a Roman villa thrown in for luck. There was only one building on the whole square. A motorway and railway sliced through the middle, but a third of the area was a country park and all the rest was open countryside. I was looking forward to roving around a pleasant landscape dotted with enormous trees.

And yet...

And yet, it turned out that the solitary building was a historic manor house that owned most of the grid square and resolutely refused to share it with plebs like me. I was shunted away from the meadows and ancient trees by signs and fences, and ushered instead down an unattractive path squashed between the motorway and a metal fence. I hoped I could at least explore a small copse, but that turned out to belong to a golf course and was also off-limits. And the lake was ringed with forbidding notices from the fishing club that owned it.

So far, the most enjoyable part of the outing had been standing on the motorway bridge and watching the hypnotic traffic hurtle beneath me.

Away from all that deserted expanse of private land, I could at least enjoy the public country park, with nice footpaths and running trails weaving through the woods. However, as it was the school holidays, the park was extremely busy with dog walkers, noisy kids and family picnics. Seeing so many of us corralled into one small slice of countryside added to my irritation. It showed how much we value being in nature, but how little of it is available. Our footpath network is brilliant if you'd like to walk somewhere, but often people just want to sit under a tree and chat or give their children space to run around and get muddy and decompress. I would love there to be more options for local people to explore on summer days than just this crowded council park in reclaimed industrial clay pits. I left the park in a huff.

As if to exacerbate my mood, I cycled over the trimmings of a flailed hawthorn hedge and got my first puncture of the year. Taking this as a sign from the gods, I conceded defeat on the day and limped to the nearest pub garden to fix my flat tyre, order a cold beer, and stew over the conundrum of land access that had reared its head surprisingly often this year.

I believe that granting wider access to anyone who wanted to enjoy and care for the land would cause negligible problems in the long run. But giving free access to absolutely *everyone* would definitely generate

some friction at first. If we could educate people to behave responsibly in the outdoors, things would be fine. Yet that is easier said than done. I understand the anxieties about unleashing more poo-bag-hangers onto an already damaged landscape before that education somehow happens.

How can we soften our inability or unwillingness to conceive of land ownership in any other than absolute terms? And how can we ensure that greater access would not further damage a landscape already blighted by unsustainable farming and fly-tipping?

One of England's challenges in increasing any rights to roam is that we have a densely packed population. Is there the capacity for all of us to access the commons? There's no doubt, for example, that when the Covid lockdowns brought more people out into the countryside, there were some negative impacts such as litter and erosion. And what if everyone went wild camping rather than paying to go on holiday?

The Lakeland charity Fix the Fells acknowledges that 'erosion from people, coupled with severe weather events and climate change, is causing ugly scars and environmental damage in the fragile mountains'. But increasing access would generally disperse people over wider areas, making the damage of tourist honeypots less of a problem. We already have 140,000 miles of footpaths, and most people (myself included) are content to use them for almost all recreational outings. And nobody sensible objects to sensitive habitats being restricted, for example during curlew nesting season.

I sipped my beer and considered how some other countries approach access, such as Sweden's policy of *allemansrätten*, which is wedded to a national pride in loving and protecting the countryside. *Allemansrätten* is the right for every man and woman to roam the countryside responsibly, and most of Scandinavia has a similar approach.*

In Estonia, 'all bodies of water that are public or designated for public use have public shore paths that are up to four metres wide… The owner may not close this path even if the private property is posted or marked with no-trespassing signs.'

Even Belarus, an authoritarian state with restricted civil liberties, insists that 'citizens have the right to stay in the forest and collect wild fruits, berries, nuts, mushrooms, other food, forest resources and medicinal plants to meet their own needs'.

Entwined with these enviable access rights is always a responsibility to respect nature and not to cause problems for other people who live, work or relax in the outdoors. The maxim is 'do not disturb, do not destroy'.

For Finns, the responsibilities are clear. 'One may not disturb the privacy of people's homes by camping too near to them or making too much noise, nor litter, drive motor vehicles off road without the landowner's permission, or fish or hunt without the relevant permits.'

Norway's policy 'is based on respect for the countryside, and all visitors are expected to show consideration for farmers and landowners, other users and the environment'.

In Iceland, 'state-owned land such as conservation areas and forestry areas are open to everyone with few exceptions. These exceptions include access during breeding seasons or during sensitive growth periods.'

Closer to home, in fact right here in the United Kingdom, the Land Reform Act of 2003 gives everyone rights of access over land and inland water throughout Scotland, subject to specific exclusions and as long as they behave responsibly. The Scottish Outdoor Access Code is based on three key principles that match the way I have tried to explore my map: respect other people's interests, care for the environment, and

**I have made a short film called* Allemansrätten, *which you can watch on YouTube.*

take responsibility for your actions.

Scotland's access rights obviously don't apply to homes, gardens, farmyards or land where crops are growing. The code clearly tells you to take your rubbish and dog poo bags home, and never to light fires or barbecues 'in dry periods or near to forests, farmland, buildings or historic sites'. If I heeded all this in England and went for a walk in my local woods, what harm would I cause?

But I do not have the right to do that. Though the freedom to roam is protected by law in several eminently sensible countries, it is regarded with horror by English authorities and landowners. (Not surprisingly, there is great historical overlap between those two groups.)

The Scottish government does a competent job of broadcasting messages about how to care for the landscape. They educate and inform visitors, meaning that fewer people cause harm, either by accident or intentionally. By contrast, England spends only £50,000 promoting the Countryside Code (at least this has increased from the £2,000 annual sum of the past decade). It is simpler just to keep folk out. But a population ignorant about how to behave in the wild world is going to create more problems than an enlightened one. And the current state of England's countryside shows that what we are doing at the moment is not working.

If we are to increase access, it needs to come with an increased responsibility towards the land. We need to have grown-up conversations that acknowledge the difficulties of trying to achieve all the contrasting things we demand of the space on our maps. The right to roam does not have to be so divisive, and it would be helpful for all parties if the discussions were more amicable. Many of us care about many of the same things, have similar values, love nature, and want to protect our wild places.

I sympathise with farmers and landowners who have to deal with littering, dogs chasing livestock, and land management guidance flip-flopping on the whim of politicians. Dog walking and biodiversity don't mix well. Essential culls of grey squirrels and deer would be harder with people rambling around. Crops have to be protected from harm. Politics, plus damage caused by irresponsible access, hinders landowners and farmers from knuckling down and delivering nature-based solutions for the public good.

While some landowners have thousands of acres, they still often struggle to stay afloat. They could cash out and sell their land, but they don't want to do that, for they value the countryside, and want to steward their patch and care for it. Over recent decades, however, this was usually done via grants that encouraged keeping the place 'tidy' and maximising food production, which we now realise stripped the soil of nutrients and the land of diversity.

Both of us appreciate this landscape – me on one side of the fence looking in, and the landowners looking out with concern about how to care for the countryside to which they've given their lives. Surely there are ways we can share both the resources and the responsibilities? I would like more access, and in return I'd be willing for more of my taxes to go towards looking after this land, both in terms of education and to support restoration work and sustainable practices.

There are lots of landowners who are itching to restore their bit of countryside once the government takes the environment seriously enough to release the brakes, set the destination, and add a little rocket fuel in the form of suitable subsidies. When that happens, they could quickly make inspirational progress.

If access to the countryside is to increase, we must each leave a positive trace on the land whenever we are in the outdoors. This means leaving places in a better condition than we find them. It is an important escalation from the long-established 'leave no trace' convention of the outdoor community. When we only leave 'no trace', nobody knows that we have been there, that we care, and that this land is important to us and to other roamers like us. A positive trace not only includes tidying up litter, but also having courteous, positive discussions with farmers and landowners about the issues surrounding rights of responsible access.

Fair access to our countryside is an important step towards reconnecting society with nature. On top of health benefits and engaging with the climate crisis, this would lead to a population that cared more and valued the important work of farmers.

'This is more than just a fence around the countryside,' goes the song 'The Commons', 'By fencing off the commons they're fencing off our minds.'

Whatever the future held, our land access issues would not be solved instantly by a frustrated and disgruntled cyclist fixing a puncture in a beer garden. But the cold pint and the sunshine certainly improved my mood.

THISTLES

'We should go forth on the shortest walk, perchance,
in the spirit of undying adventure.'

Henry David Thoreau

The gate's clang startled a buzzard who lumbered off the ground and flew into the sanctuary of the trees. I stood still in the field, feeling myself beginning to slow down and unwind. I breathed in the smell of hay, blinked at the sunshine, and reminded myself that things couldn't be too bad if I got to call this 'work'.

Riding here had been a confusing maze of winding lanes and high hedges, so I hadn't yet orientated myself with any other familiar grid squares nearby. The road had been too narrow for cars to pass my bike safely, so I'd had to stop and tuck in whenever a vehicle appeared. This allowed me the chance for a blackberry update, nibbling one or two while I waited for each car to pass. A few were ripe and swollen, but most were still small green nubbins.

I listened to the whine of motorbikes on a distant race track, the sounds carrying on the soft breeze and highlighting how quiet this square was. I felt distracted today, uncertain as to what I should do after this project. That day was approaching, but I still had no answers.

My feelings had morphed over the course of the year, from realising there was more of interest on my map than I'd given it credit for, through a sense of slowing down and connecting with local nature, and on to accepting my life more and being grateful for all that I have. But this was also mixed with an increasing alarm over the state of the environment and a desire for more people to wake up and care. What could I do about this?

I've always had plans and goals throughout my adult life, so it was disconcerting and disorientating to be uncertain about my future direction. But I tried to shrug off those worries for now, to just appreciate where I was today. I saw a long valley dropping away towards a grid square where I remembered drinking hot soup back in the winter. The view settled my mind. I knew where I was.

I left the fields and entered a wood because my map suggested there was a pond in there. But barbed wire and brambles kept me away. Google Maps' satellite mode consoled me that the pond was just a brown, dried-up affair, but I don't know for sure. For, if truth be told, I did not try particularly hard to reach it. After ten months of sneaking around woods, leaping fences and pushing through undergrowth, I was tired of how often a stroll in a wood became illegal trespass, and how unfair it is that being a white middle-class bloke meant I found all this less intimidating than many other groups might do in our countryside.

So today I didn't fight it, but just walked on through the wood to another farm. Every field here was covered in thistles, waist high, prickly, and choking all other growth. While all plants are important to nature and wildlife, a few can get out of control if nature's balance is thrown off kilter.

Species such as rushes, ragwort and thistles can rapidly out-compete other plants, so are categorised as 'injurious weeds', making it the landowner's responsibility to control them. It is possible to find a workable balance for dealing with them using methods that do little harm to other wildlife and can help to increase biodiversity. Thistles are valuable for wildlife, and adored by one of my favourite birds. Goldfinches depend on their seeds, as do greenfinches, siskins and redpolls. Painted lady butterflies eat the thistle's leaves, while bees and white letter hairstreak, peacock, and meadow brown butterflies rely on the flowers for nectar. Many insects also overwinter on the thorny stems.

True to form, a charm of goldfinches – a hundred or more – burst from the thistle field with their looping flight. I had never seen so many. The goldfinch is a charming bird with a red face, white throat and black and yellow wing patches. Ever since I started hanging seeds outside my shed, they have been regular, noisy, colourful visitors, keeping me company with my daily writing procrastination. They are sociable and boisterous, twittering and wittering all day long. Because they are associated with thistles, goldfinches used to feature in Christian symbolism for the Passion and Jesus's crown of thorns. A little goldfinch in paintings could represent Jesus and Mary's foreknowledge of his crucifixion.

Until the 19th century, thousands of goldfinches were caged and sold each year for people to enjoy a shard of colour and song in their homes. It sounds a sad sort of pet to me. A goldfinch also offered a rare glimpse of beauty for the old Irish farmer in Patrick Kavanagh's poem 'The Great Hunter'.

'The goldfinches on the railway paling were worth looking at –
A man might imagine then
Himself in Brazil and these birds the birds of paradise,
And the Amazon and the romance traced on the school map lived again.'

The profligacy of the thistles' wind dispersal was extraordinary in the fields I walked through. Thistledown covered the ground like snow and the air was full of seeds drifting on the breeze. Because wind-dispersing plants have no control over where their seeds will land, they have to produce vast numbers and then put their faith in a seed. *Faith in a Seed*, incidentally, was the title of Thoreau's final manuscript, a study of wind dispersal that not surprisingly didn't gain the same level of fame as *Walden*, a tempting tale of turning your back on the frustrations of modern society to live in a cabin in the woods.

Scarlet pimpernel grew in the field margins, a once common wildflower now in decline because of intensive agriculture. It is sometimes known as old man's weathervane because the petals close when atmospheric pressure falls. The humble flower became the emblem of the eponymous, chivalrous hero who operated in disguise, rescued aristocrats from the guillotine, and signed his messages with the pimpernel.

I passed a man walking his dog, tennis ball in mouth (the dog, not the man). One said hello (the man, not the dog) and the other wagged

his tail. A teenage son trailed by several yards, glued to his phone with headphones in. He didn't notice me.

I climbed a stile and entered an overgrown wood where I found a large memorial stone, freshly positioned and strewn with roses. It was a tall, uneven boulder, almost my height, and the first memorial I had found away from a churchyard. Getting it here had been an achievement, for it weighed a couple of tonnes, at least. There were no words on the memorial, just an engraved star in a circle – a pentagram – which was originally a symbol representing the five wounds of the crucified Christ (four nails, plus a spear in the side) and regarded as a deterrent against evil.

There was no message on the roses propped against the stone, so I didn't know whose memorial it was. Then I spotted, tossed into the undergrowth, the foam outlines used at funerals to spell out names in flowers. The green shapes edged with red ribbon were my only clue towards the life commemorated here: MUM and NAN.

On the far side of the wood, I relaxed on a bench decorated with a motif saying 'Live, Love, Laugh & Be Happy'. Earlier I'd passed a sign on a farm gate with a picture of a cow and the words 'Slow down'. It was intended to caution drivers about the milking herd, I imagine, but I had chosen to see the wise cow advising me to slow down, to chew the cud and enjoy the sunshine. It was the cow that prompted me to

pause on this bench.

I liked it up here at the head of the valley. It was a lofty, quiet backwater with a web of lanes leading nowhere much. I'd never taken much interest in this area before, even when out exploring on my bike. But I had discovered today that it was a new approach to a valley I was fond of and I appreciated getting a fresh view all the way down to the flat lands on the edge of my map. The tall buildings of the city sparkled in the crisp summer light. I thought of all the busy, ambitious people there. They looked close, but right now felt very far away. I might have been unclear about my life once this project ended, but sitting for a while on this bench seemed as good a way as any to figure things out.

'Live, love, laugh and be happy,' it advised.

'Slow down,' added the cow.

BLACKBERRIES

'I remember how glad I was when I was kept from school a half a day to pick huckleberries on a neighbouring hill, all by myself, to make a pudding for the family dinner. Ah, they got nothing but the pudding, but I got invaluable experience beside. A half a day of liberty like that was like the promise of life eternal.'

Henry David Thoreau

Today's grid square was a rare outing to the far side of the river, to the very edge of the map itself. It felt like a new country. Over that next hill lay lands unknown, and maybe even dragons.

I cycled up a stony bridleway through a wood, making sure to savour the greenness before the leaves fell for another year, to store away the memories as nourishment to get me through the winter. The year was winding round to its close, and I was going to miss these outings. They always cheered me up after tedious bouts of real life, such as queuing this morning to collect a parcel from the post office, which turned out to be in some other distant depot. Holly berries ripened in the dim woodland light. The path became a holloway, with beech trees arching overhead and their tangled roots exposed on the elevated track sides. A nuthatch scurried up and down a trunk, calling 'dwip, dwip' as

it searched for food, then hung upside down while it ate.

At the top of the woods, I found a lookout point and a picnic table bearing the scorch marks of half a dozen disposable barbecues. There were long views down over the old village and the new development on its flank, then over the river and east towards the wooded ridge I'd visited often this year.

Dropping back towards the village, I freewheeled through fields of mown straw gathered into rows to dry before being baled. It was September now, the season of sleepy-eyed kids slumped at early bus stops in new school shoes and too-large blazers. A grey haze of coolness hung in the air with a faint scent of honeysuckle. I wished I had a woolly hat, even as summer's swallows and martins still circled overhead. One day soon, I would notice that I hadn't noticed them in a while, and they'd have all gone and left me, flown south to African skies without even saying goodbye, like the best years of my life if I don't make the most of every day.*

Someone had repurposed the obsolete village telephone box as a library, which I always like to see. I glanced inside, secretly hoping I might see one of my own books in there. I didn't, of course: I rarely see

**A visitor to Ikkyū, the renowned 15th-century Japanese Zen monk, asked him for some wisdom.*
'Attention,' Ikkyū offered.
The disappointed visitor pushed for more.
'Attention, attention,' expanded the monk.

them even in bookshops, let alone in a phone box of discarded best-sellers and bodice rippers. Still, a man has to dream…*

The narrow road through the village couldn't cope with the busy back-to-school traffic, as both sides of the street were already lined with parked cars. I slipped past the chaos along the pavement, imagining how brilliant it would be if all those youngsters cycled the short journey to school with their friends instead.

A sign on the church invited me in, 'You are welcome to share the quiet, beauty and friendship of this church.' I passed beneath the lych-gate, built to commemorate the local men killed in the First World War. In the graveyard, I found the tomb of the oldest man aboard the *HMS Victory* at the Battle of Trafalgar. He was in his late sixties on the day Lord Nelson died.

Alongside was the Victorian grave of a woman who had twenty-two children, 'eleven of whom are buried near this spot'. And close by was the modern grave of an infant who died the day she was born, decorated recently with a wreath for what would have been her 18th birthday, from 'Mummy and Daddy'. It brought to mind Wordsworth's aching tribute after losing his six-year-old son, 'I loved the Boy with

**While in confessional mode, I admit I sometimes dream of seeing somebody enjoying my books on a train. I sit nonchalantly and anonymously nearby, watching their rapt expression or hearing hoots of laughter as they tell fellow passengers what an underrated genius the author is. There: I told you I was a vain and deluded fellow!*

the utmost love of which my soul is capable, and he is taken from me – yet in the agony of my spirit in surrendering such a treasure, I feel a thousand times richer than if I had never possessed it.'

Leaving the churchyard, I ducked into a small wood to dodge a rain shower. Dirt bikes had carved an ugly but fun racing course around a tall stand of oaks. Slender silver birches had bracelets of green moss around their bases. I liked the atmosphere in the wood. It is interesting how some woods just feel right in intangible ways.

I smelt weed and spotted a pair of lurcher dogs running through the trees. Soon after, a man in a flat cap appeared, smoking a joint. His teenage son had gone heavy on the Lynx deodorant. The pair were carrying air rifles and scanning the tree canopy. We chatted about the rain, as Brits do, and then I asked what they were hoping to shoot.

'Squirrels, pigeons, anything really. But it's raining, innit? So they're all like "fuck you" and hiding, know what I mean? What about you?'

'A bit like you guys,' I answered. 'Just looking for anything interesting. I shoot with my camera, though.'

I brandished my camera, and they laughed and wished me luck. A few minutes later I heard the pop pop pop of pellets, the yapping excitement of the dogs, and some squirrel or pigeon had run out of luck.

The river lapped against the margins of the village, muddy and full. I listened to the susurration of the wind whispering in the reeds, then doubted myself whether susurration was even a real word. (It turned out that it is real, from the Latin for 'whisper', and was apparently Sir Terry Pratchett's favourite word.)

I followed the river downstream to the edge of my grid square. I passed pylons in the reedbeds, circled with barbed wire and signs warning of the 33,000 volts fizzing through the cables. I passed a smelly sewage works and a smelly cattery. I continued past a tribute to some lime-kiln workers who died here, past the site of the cement factory that once brought wealth to the area, and past where a ferry crossed back and forth for hundreds of years until the 1960s. You could take a bike on the ferry, but it cost a halfpenny extra. And I paused on a quiet bench to watch the water pass for a while and imagine all these separate pasts.

Peak blackberry season had arrived at last, and with that, my official onset of autumn. I gorged on a large patch, my body an efficient eating machine whirring through the movements of pick, eat, pick, eat. My eyes scanned the brambles for the next ripe berry as my fingers automatically found my mouth. Blackberries must be the highlight of the forager's calendar (except for those savvy enough to harvest wild mushrooms). They are easy to find, plentiful, won't kill you, and actually taste delicious, unlike a lot of foraged food which just tastes of leaves.

Humans have enjoyed blackberries for aeons. The Haraldskær Woman, whose superbly preserved 2,500-year-old body was found in a Danish bog, had blackberries in her stomach. *The London Pharmacopoeia* of 1696 listed all permitted medicines and remedies, and they included blackberry wine and cordial. Blackberries were also believed to defend against certain spells, providing, of course, that they were gathered during the correct phase of the moon. You could crawl through bramble patches to cure boils, they said, or even to repair a child's hernia. Tradition cautioned that blackberries should not be picked after 10 October, Old Michaelmas Day, for that was when Lucifer was banished from heaven and fell to earth, landing in a blackberry bush. Brambles were planted on graves to deter grazing sheep, but also because they helped keep the dead in.

Twenty years ago, mention of the word 'blackberry' conjured thoughts of the wildly successful phone company loved by Barack Obama and a generation of businessmen and women. Yet BlackBerry, the brand, has disappeared from the sector and our consciousness. Whether you are a blackberry plant or a BlackBerry phone, the difference between survival and extinction is about adapting to changing environments. BlackBerry phones are now gone and forgotten. They didn't adapt to cope with the environment or the competition, so became as extinct as the passenger pigeon.*

In 2007, a new edition of the *Oxford Junior Dictionary* included such new words as broadband, but removed many describing the natural world, such as heron and lark. It was done in response to the current frequency of words in the daily language of children, which felt sad and alarming. An open letter signed by naturalists and artists disputed

**The iPod was conceived as a way to sell more computers, but it became a triumph in itself. Yet Steve Jobs realised phones would soon be able to store as much music as an iPod. So he led the pivot from iPod to iPhone before it became obsolete. The rest is history.*

the decision, pointing out that 'there is a shocking, proven connection between the decline in natural play and the decline in children's well-being'.

The philosopher A.J. Ayer raised the notion of logical positivism, that if we don't have a word for something, we cannot think of it, let alone care for it. The direct connection between our imagination, our ideas and our vocabulary had certainly been true for me this year.

Unless we have a word for something, we cannot conceive of it. Unless we explore our neighbourhood, we can't imagine what might be right under our noses, nor be able to celebrate it, mourn its demise, or take action.

Summer was ebbing away and the swallows would soon fly south. My year exploring this map was drawing to a close. But autumn's glorious colours were coming and the hedgerows were full of fruit. It would be up to me to choose what happened after this journey finished, and then to embrace that. There is nothing permanent except change.

SATISFACTION

'Watching ourselves and watching the world are not in opposition; by observing the forest, I have come to see myself more clearly.'

David George Haskell

Though the silver birch trees were turning to autumnal gold, summer was back this week with a fury, despite me writing it off. But it was probably too early to speak of an Indian summer. The earliest known use of that phrase comes from a Frenchman called John de Crevecoeur in the eastern United States in 1778. It perhaps referred to a spell of warm weather that allowed the Native Americans to continue hunting a little longer. The phrase reached Britain in the 19th century, replacing 'Saint Martin's summer' that had been used to describe fine weather close to St Martin's Day on 11 November.

The sun was hot on my dark T-shirt, and I pulled my cap down to shade my eyes. As I rode from street to street towards the edge of the small town, the houses became larger and more spaced out, until I reached the land of Waitrose delivery vans and lawns with willow trees. There was frequently a substantial differentiation of wealth within the grid squares I visited, and there is a depressing correlation between health and longevity. I wondered if there is also a corresponding link between wealth and other measures such as happiness.

Research suggests that half our happiness is down to genetics and personality, and half comes from the other factors in our lives: work, health and relationships. Happiness is linked to income, but only up to the point of living a comfortable lifestyle. Once people earn above the level of having to worry about money, happiness levels plateau. How much we earn compared to those around us affects our happiness a lot more than the actual sums involved. Gardeners and florists are the happiest of us. Self-employed workers are more content than employees, while bankers and IT workers are the unhappiest, though not the poorest.

I felt my spirits rising as I rode away from the streets in search of countryside. A sunny day always lifts my mood. People in warm countries are usually more satisfied with life than those in cold and miserable climates (with the exception of the Scandinavian countries who always seem to score well in any poll about doing life well). But the increased pleasure of sunshine only has a short-term influence on overall life satisfaction, because the curse of hedonic adaptation means we soon take all good things for granted, even endless sunny days. In the end, the best predictor of how satisfied we will be with our lives in the future is how satisfied we are with our lives today.

I certainly felt satisfied, riding through the countryside in the sunshine. This project had emphasised to me that getting more people out into more of the countryside more often would have many benefits. This notion is epitomised by the Nordic word *friluftsliv*, or 'fresh air life'. It is enshrined in Norway's Outdoor Activities Act so that the 'opportunity to practise outdoor life as a health-promoting, well-being-creating and environmentally friendly leisure activity is preserved and promoted'.

Friluftsliv is a way of life taught to every Norwegian child, and you can even study it at university. Its principles are also perfect for exploring a local map. They include connecting by sharing your explorations with others or meeting people on your map, observing and being curious, learning, and giving your time or expertise to leave your map in a better place.

A car accelerated past me as soon as it left the restricted 30mph zone of the village. The driver gunned his engine at the speed limit sign and screeched away round the corner. Fast driving in rural areas and a lack of cycle lanes or pavements is a significant barrier to encouraging more people to walk or cycle in the countryside.

I was glad to get off the road and head down a footpath alongside a well-fenced, almost fortified chicken coop. A piratical Jolly Roger flag flew from the roof, reinforcing the siege mentality of those chickens against the perennial risk of foxes. The name Jolly Roger came from the French *joli rouge*, or 'pretty red', because pirate flags were originally blood red as a warning that no mercy would be shown if a ship opted for battle rather than surrender. Pirates adopted the skull and cross-bones from the symbol used in ship's logs to represent a death on board. Since the decline of piracy, other military units have adopted the Jolly Roger's insignia and spirit, including US Naval Aviation squadrons, Chilean paramilitaries and Portuguese Lancers.*

I paused along the path to feast on blackberries again. It was a popular footpath so most of the low-hanging fruit had already gone. But a benefit of being above-average height was that I could reach some higher berries, and by standing on the frame of my bike I could stretch even farther. As I foraged, I saw that acorns were beginning to grow and swell on the oak tree nearby. 'I hold in my hand not a single tree, but a community-to-be, a world-in-waiting,' wrote Robert Macfarlane, describing an acorn. Up to 2,300 species live in an English oak: 716 lichens, 108 fungi, 1,178 invertebrates, and many owls, bats, birds and butterflies.

Breakfast over, I crossed a railway line on an exposed crossing where you had to walk over the tracks, carefully looking left and right before setting off. A kestrel was hunting its own breakfast, hovering in the warm sky in a picture of taut concentration. The blackthorn hedges, so frothy with lacy white flowers back in the spring, were now heavy

**Admiral Sir Arthur Wilson VC, Controller of the Royal Navy, once complained that submarines were 'underhand, unfair, and damned un-English'. He sniffed that we should 'treat all submarines as pirates in wartime... and hang all crews'. In rebellious response, submariner Lieutenant Commander Max Horton flew the Jolly Roger after sinking two German ships in 1914. It became common practice for submarines in the Second World War to do this after a successful mission, and the Jolly Roger has since become the official emblem of the Royal Navy Submarine Service, accompanying their motto of 'We Come Unseen'.*

with blue-black sloes. A fruitful autumn of picking sloes depends on the spring and summer weather. A wet and cold year means no sloes, but too many hot and dry days make small, shrivelled fruit. It is a fussy, Goldilocks type of fruit.

A chiffchaff chirped from deep within the blackthorn's spiky sanctuary, and a tiny stream trickled beneath the hedge, making me regret how few streams there were on my map. Today, of all days, I longed for cool water to submerge myself in. It was scorching.

Fortunately, today's square had a lake marked on the map. I turned towards it, eager to swim. A footpath ran towards it for hundreds of metres across the grid square. I'd assumed this would be an opportunity to enjoy the countryside, but it turned out to be a gloomy tunnel hemmed between high house fences, a quarry fence and overhead trees. The houses formed the outskirts of a posh town just beyond the reach of my map. Perhaps this street was not so fancy, though: a sticker attached to a road sign said, 'this sign has no scrap value'. The leaves on the maple tree above the traffic lights were beginning to turn from green to amber and red. Autumn's colours were beginning to dominate.

Unfortunately, not only the lake but more than half the grid square was cordoned off for a sand quarry, and I had no intention of climbing into a working quarry. As the signs posted along the fence warned, 'people are killed and seriously injured every year in quarries'. I wanted to get at least a telephoto glimpse of the lake and the quarry in order to better understand today's square, so I set off on a big loop to search

for a viewpoint.

I found a public footpath that ran right through the working part of the quarry. Much as I'm a champion for land access, this seemed like a ridiculous place for a path, with all sorts of dangerous machinery roaring around. Massive diggers were literally removing a hillside, and I felt like a little boy as I enjoyed watching them dump their loads onto long conveyor belts to be carried away to build distant roads and towns.

Lying still amidst the noise and dust and apocalyptic vehicles was the tantalisingly blue water of an old sand pit. Tempting for swimming, but acceptably off-limits for me. I settled for cycling to a nearby village to search for some shade.

The church there was unlocked and the cool gloom and muffled peace was a respite from the scalding, glaring day outside, much like swimming underwater. Stained glass windows gleamed as the sunlight poured burnished pools of colour onto the flagstone floor. Those images must have appeared so magical to devout and simple congregations before our sense of wonder was numbed by HD TVs and 4K phone screens dishing up dazzling colour every hour of the day.

One thing I always appreciate in ancient churches is the sheer bloody oldness of Britain. People have sought rest, respite, solace and perspective on this very spot for thirty generations or more. That felt meaningful and important to me, despite not believing in any god. The

sands of time trickle gently here. Slow down. Today is enough.

A marble plaque on the church wall listed benefactions gifted to the local community centuries ago. 'John Porter devised an annuity of two Pounds to be given to the two oldest married persons, at the discretion of the Minister. William Baker bequeathed forever twenty-six shillings yearly to be paid to the oldest deserving poor person. And the Lord of the Manor agrees to give annually 500 fagots [a bundle of sticks bound as fuel] to the poor of the parish.'

I wondered how long those pledges lasted for. With the powers of cumulative interest, they would be worth a fortune by now if they'd been invested in index funds. Rather than the ever-growing financial and carbon debt that we are passing down to our children, what useful legacy should our generation leave for the future?

There was a box on a shelf filled with 'Quarantined Prayer Books' from the great medical drama of our age. That, in its turn, will be replaced by the next big thing. I looked at the grave set into the floor by the altar, belonging to a sombre-looking knight. He was buried here one September day, 626 years ago. I wondered what he had enjoyed doing on sunny September days like today, beyond rescuing damsels and slaying dragons. What did he worry about? What did he do with five minutes to spare without Twitter to look at? The well-worn tombs on the flagstones and the fading names of once-prominent personalities reminded me not to fret about chasing fame or fortune. Everything I do will be forgotten sooner rather than later. In other words, go and enjoy the sunshine.

And with that, I headed out into the heat in search of a decent cup of coffee. I enjoyed a flat white in the sunshine while contemplating the meaning of life, before scrolling around the maps on my phone, feeling I had been rather premature in declaring the end of summer five weeks ago. Happiness might well be tied to income and personality, but it's surely also connected to finding a river to swim in on a hot September afternoon.

BEES

'Live in each season as it passes; breathe the air, drink the drink, taste the fruit, and resign yourself to the influences of each.'

Henry David Thoreau

A long row of black poplar trees escorted my road towards the low horizon. I passed a row of small industrial units, then a house offering rosy windfall apples and pears in a chipped, white ceramic bowl on the doorstep. Voices carried from an open upstairs window, engrossed in a Zoom call about something or other.

A cluster of beehives stood in the corner of a field. The coming cool weather would soon quieten the hives, but today the sun was warm and the bees were busy. They fly tremendous distances, racking up round trips of up to ten miles to forage for food. Each jar of honey contains nectar from two million flowers, with a corresponding flight distance of 90,000 miles, or more than three laps of the planet. Yet each bee produces only a twelfth of a teaspoon of honey in its life. So it is an extraordinary team effort that depends upon bees sharing information about the food sources they find, by 'dancing' for each other.

The waggle dance involves flying in a straight line to show the direction of the food relative to the sun, then performing a series of

loops related to the flowers' quality. The bee also beats its wings and waggles its abdomen to create vibrations that give extra information about the nectar and pollen's location.

Bees are cooperative, communicative insects, complete with solar compasses, inbuilt clocks, the ability to communicate with plants via electric signals, and a sting in the tail. They pollinate most of our wildflowers and many important crops. Bees are amazing. But after 100 million years, they are now at risk as we kamikaze towards 'insectageddon' and the extinction of up to 70 percent of our wild species.

One of the many problems bees face is fragmentation of the routes they use to move around. The splintering of wild places makes it challenging for wildlife to move across the countryside. Imagine trying to travel without roads or train tracks. The Buglife charity is establishing 'B-Lines' to help, a series of insect pathways with 'stepping stones' of wildflower habitats along the way, like service stations for bees. B-Lines will benefit many other types of wildlife too, if farmers, landowners, local authorities and gardeners cooperate to create enough of them.

I cycled past a closed-down pub and a 200-year-old obelisk boundary marker, then followed a stretch of canal choked with bullrushes. The water was clear and teeming with life around the heaps of dumped furniture and tyres. A red-beaked moorhen sat on her nest, like a queen on a throne amid the junk.

Somewhere in the distance, a dog barked, a man shouted instructions, and an old tractor chug-chugged through a field. Two collared doves watched me ride by from their perch on a phone wire. A recovery vehicle overtook me, the driver texting with one hand and holding his coffee in the other. Horses grazed in a paddock as a police forensics van drove by. Someone had chucked a kids' slide halfway over the hedge. There was a sheen of dew across the fields as summer continued blending into autumn. The verges were messy with litter, but there were also clusters of wild sweet peas strung with spider webs.

For most of the year, we don't particularly notice spiders. But when they search for mates in the autumn, some species spin webs that can cover entire fields in shimmering sheets of silk. Silk is a watery protein gel that only solidifies when it is stretched. There are 650 species of spider in Britain, of which half are the tiny money spiders who weave enormous webs. Superstition says that if you find a money spider in your hair, you'll soon come into money. To make extra sure of the good fortune, spin the spider three times around your head by its web before releasing it.

Most spiders adopt a specialised niche, such as reedbeds, dunes or heathland. Their webs are similarly diverse and can help identify which spider family made it. They tend to make one of seven styles: orb, sheet, tangle, funnel, lace, radial or purse. All spiders make silk, but they don't all create webs. They have several silk glands – spinnerets – which produce different kinds of silk for different tasks, including releasing draglines into the wind to balloon up into the air, building shelters, courtship, and, of course, for creating webs to catch prey.

A typical web takes a couple of hours to construct. The spider begins by allowing a silk line to drift in the breeze to bridge a gap. It strengthens this strand with extra threads, then adds the radial and spiral lines. Finally, it tidies up its work by removing all the knots from the middle of the web and replacing them with a silk lattice.

My map showed an orchard north of the village, but that was gone, replaced by harvested fields of bales that were giving off a ticking sound, like bubble bath popping, as damp mould in the hay respired, releasing moisture and heat. The air smelt sweet, that earthy scent of the last days of summer. I cruised down the middle of deserted roads, slaloming past rustling poplars. I realised that I was dawdling, basking in the warm weather and recharging ready for the dark months that would return soon.

The road ran out in a tiny hamlet of thatched cottages clustered around a small church, like a small island among a sea of flat arable fields. It felt like the end of the world, or at least the end of my map. The wide-open marshes stretched beyond the red-tiled church with its striped walls of ragstone and knapped flint.

I propped my bike against a gravestone and pushed the church's magnificent 600-year-old wooden door. It was weather beaten, knotted, and carved with birds, flowers and faces. I wondered who the carver had based those faces upon. The keyhole was three inches large, but the door was unlocked and swung open. I stepped into the cool, quiet building and allowed my eyes to adjust to the gloom before absorbing the ancient atmosphere of Norman worshippers and Benedictine nuns.

Leaving the church, I followed a footpath towards the marshes. Progress was slow as I kept stopping to eat blackberries. There were thousands upon thousands. Even the fox poo was stuffed with berries (I didn't eat that). I soon became more discerning in my fruit selection: the berries on the shaded north side of the path were nicely chilled, but not as plump as those on the southern side. As the noted botanist Louis Armstrong sang, '*life can be so sweet on the sunny side of the street*'.

A woodpecker flapped indignantly into the trees as I poked at a blood-red fungus on the trunk of an old oak. If you slice through the flesh of the beefsteak fungus, it resembles a juicy steak though sadly is not as tasty. Young specimens are edible, however, and taste best when slowly simmered. Furniture makers prize oak timber that has been infected by the parasite, referring to it as 'brown oak' and valuing its dark hue.

The fields became scruffier as I got nearer to the marsh, increasingly unkempt and close to reverting to wildness. It was a jumbled, broken down and mutable landscape. The grass was tussocky, like bed head, with bushes, brambles and oak saplings dotted around. There was a left-alone vibe here that I enjoyed, although if left completely alone, the marshland would soon be swallowed by scrub and woodland. The countryside requires a gentle helping hand to preserve the variety of its habitats.

This corner of my map felt eternal and unchanging, but even out here, development marched on. A rough handwritten sign nailed to a fence post complained about the 'Development of 6 houses proposed

here = light polution [sic] noise extra trafic [sic], No gas, mains drains no street lights no paths. Small country lane. Please object.'

On my way back to where I'd begun today's ride, I came across an apple tree growing wild by the road, with a cluster of red fruit near the top. I gave the trunk a bit of a shake and caught my first wild apple of the season. As I crunched the tart, white flesh, I remembered eating an apple like this back on my very first grid square. Over the course of this year, I had learnt so much. Not least of all was realising the power and wonder of opening my eyes and looking around. I had been walking around in blinkers all my life.

Thoreau, a man better tuned than most to the simple multitudes of local living, was a fan of wild apples. 'These apples have hung in the wind and frost and rain till they have absorbed the qualities of the weather or season, and thus are highly seasoned,' he wrote. 'And they pierce and sting and permeate us with their spirit. They must be eaten in season, accordingly – that is, out of doors.'

Yet he also conceded that they can sometimes be 'sour enough to set a squirrel's teeth on edge and make a jay scream'.

The Welsh word *cynefin* encompasses the multiple, tangled threads in our landscapes and experiences that influence us deeply but are hard to put your finger on. It combines the rootedness of where you live with the relationship you have with that land. These fields and paths and streets had shown me a great deal, but they also asked hard questions. Who are you? What are you doing here? What matters to you? What are you going to do about it?

And all the while, the seasons roll round and apples grow and fall.

VALLEY

'Nature does not need to be cleansed of human artefacts to be beautiful or coherent. Yes, we should be less greedy, untidy, wasteful, and shortsighted. But let us not turn responsibility into self-hatred. Our biggest failing is, after all, lack of compassion for the world. Including ourselves.'

David George Haskell

Walking always feels very different from the running and cycling I usually do for exercise. I'm generally too impatient to walk somewhere if I could run or ride instead. But the way I think changes depending on my mode of transport. Slow my legs and my mind starts to slow too. When you walk, you can stop at any time to poke something with a stick, make a note or take a photo. Walking is a movement that invites stillness. So I decided to walk this week for some deliberate slowness.

I got a positive feeling about today's grid square as soon as I arrived. On previous outings, I had often looked in this direction and thought, 'It looks nice over that way.' The omens were promising, with plenty of contour lines and no roads.

There was also the bonus of the sun blazing away again, like a late September gift. This week marked the autumn equinox, and the leaves were just starting to change colour. It was a beautiful season and, for once, I fully appreciated the here and now. The kind weather won't be here much longer. Enjoy it. These light mornings, this honeyed sunshine will not last. Make the most of it. These were fine days. I needed only to stand still and look around to have enough.

I was walking through actual, proper countryside towards an actual, proper hill, woefully scarce in this flat corner of the world. A very old man shuffled towards me on the footpath and I stepped aside to allow him to pass. He acknowledged me with a slight nod of his head. His body was frail and stooped. He was short of breath and unsteady on the uneven ground. The skin on his face had sagged and his eyes were red-rimmed and sunken. But I was probably as close to his age now as I was to being a kid running around the playground. The seasons only turn forwards. Don't take my fitness for granted. Don't take these days for granted.

This was a season of abundance, with red berries on a whitebeam tree, clusters of black fruits on a buckthorn, and spiky hawthorn hedges filled with red berries known as haws, an important winter food source for hawthorn shield bugs, yellowhammers and migrating birds such as redwings, waxwings and fieldfares.

The 'chuck-chuck' of a pheasant's alarm call caused me to look up, and sure enough I spotted a fox in the lee of a hedge, slinking up the valley, its mind set on mischief in the soft sunlight. The morning felt

as though it was waiting for something to happen. This little map had grown on me a lot this year.

A band of jays squabbled in the treetops as I followed a shaded footpath and discovered a monument hidden up an overgrown path. More weeks than not, my grid squares kept throwing up quirky surprises that prompted further internet explorations back home. This memorial was for an aviation pioneer who used the valley's steep slopes to test his prototype gliders at the end of the 19th century, including one that set a world distance record. Without the inconvenience of dying in a crash landing setting him back somewhat, he might even have beaten the Wright brothers to the glorious immortality of powered flight.

This valley felt undiscovered and wild. It was a mixture of chalk grassland with scrub turning to woodland at the top of the valley. What I casually call 'scrub' actually consists of any number of species about which I don't know much. I put in few minutes of effort with my Seek app and found I was among wild carrot, devil's bit scabious, strawberry clover, agrimony, wild basil, thyme and dog rose, not to mention all the assorted grasses that glimmered with dew and spiders' webs, home to chirping and hopping insects. This briefest of efforts at observing and learning had made the land come more alive, and I was already asking myself, 'Could this be my favourite grid square of the year?'

This grid square was a good example of the buzzword 'rewilding', the process of restoring ecosystems by allowing natural processes to take over and form stable habitats. Rewilding increases biodiversity, captures carbon, and costs little. It can involve giving nature a helping hand by rewiggling rivers, blocking drainage ditches, planting the right trees in the right places, reintroducing missing species and addressing the harmful effects of overgrazing. It can also involve as little as stepping back from the land and leaving nature to heal itself.

Britain is one of the most nature-depleted countries on the planet, yet we can all do something about that simply by giving back some of our gardens, verges, playgrounds, churchyards, and farmland to nature. Rewilding is simple, affordable, surprisingly speedy – and it works.

Britain has just 13 percent forest cover, compared with a European average of 37 percent, and much of what we do have is commercial monoculture plantations with little wildlife. If the best time to plant a tree is twenty years ago, the second-best time is right now. Anyone can plant a tree. You might not see it fully grown, but your grandchildren will appreciate it.

This notion of doing something to benefit later generations was vital to the 'seven generation stewardship' concept of the Haudenosaunee confederacy of Native Americans. Its essence was that 'in every deliberation, we must consider the impact on the seventh generation... even if it requires having skin as thick as the bark of a pine'. It is a generous approach that counters our own culture of 'get rich and buy stuff. Do what we want and burn the place down on the way out.' Considering how our actions will affect our descendants seven generations from now might change that, as well as making us happier and healthier.

Opponents of rewilding mark folk like me down as deluded hippies. They assume rewilding means reducing food security, removing people from their homes, and that farmers will go bust, along with all those careers connected to them, plus rural shops, pubs and schools. But rewilding can boost employment in rural communities through job diversification, education, and nature-based tourism. It can improve farm yields by providing habitats for pollinating insects, restoring the soil, and mitigating drought and flooding. Even intensive farms can improve their hedgerows and have buffer strips around fields to help wildlife and reduce run-off. It is not a question of rewilding *or* farming.

Rewilding is not an attempt to hark back to an unrealistic, idyllic prior state, but a way to allow the countryside to become wild and diverse again alongside the demands of our 21st-century needs. We *can* balance producing food, keeping farmers on the land and increasing biodiversity, but everyone will be worse off if rewilding becomes a battle between farmers and conservationists. We need to act swiftly, decisively and ambitiously if 30 percent of our land and sea is to be restored for nature by 2030 – a government commitment on which very little progress has been made.*

Sceptics worry about losing what they call traditional ways of life through rewilding, all for the sake of a few trees and birds. Yet this is a shifting baseline perspective that inadvertently defends the intensive,

**If the Royal Family, Ministry of Defence, Church, and National Trust also got involved in restoring the landscape, that could add up to 823,051 hectares to the rewilding conversation, more than 8,000 of the grid squares I have been exploring on my map.*

unsustainable changes in land use since the Second World War. It is part of the 'kalcidoscope myth' against which the ecologist Oliver Rackham cautioned, our fondness for assuming that landscapes never change. Farms traditionally used to be diverse, wild, nature-friendly environments, and they can be again.

Rewilding the least productive fifth of our farmland to help nature recover would reduce current food production by less than 3 percent. That could easily be made up by increasing efficiency elsewhere and making minor adjustments to our diet. For example, swapping some nature-depleted fields full of livestock for one field of crops and an area of rewilding balances in terms of calories and protein produced for us to eat.*

While I often point the finger at livestock for many of our countryside's problems, this 'natural' valley only looked the way it did through grazing management. Allowing a few cows occasionally to graze here prevented the land from being engulfed by brambles and helped the ecologically valuable grasslands to flourish. Pasture-fed animals, in small numbers, play a role both in the future of sustainable farming and a rewilded landscape filled with life.**

For thousands of years, farmers have had one vital mission: to produce as much food as possible. That focus now needs to shift towards producing what a farm can generate naturally, striking a balance between commercial and environmental sustainability, as well as using corporate investment in carbon credits to rewild spare land.

**You need 119.49 m^2 to produce 1,000 kilocalories of beef and 184.8 m^2 for 100 grams of protein, compared with 1.44 m^2 and 4.6 m^2 respectively with legumes and pulses.*

***Britain's uplands, where I grew up, are picturesque, treasured, sentimental, and culturally important landscapes. But they are also terribly nature-depleted. Many upland sheep farms would also lose money every year were it not for subsidies. They produce a tiny proportion of our food and have negligible commercial impact.*
Nevertheless, we do not want to lose those upland farms, even though sheep, in particular, wreak havoc with biodiversity unless their numbers are kept very low. Why not pay those farmers to restore the landscape as well as producing food? Rewilding and regenerative agriculture are essential allies for our future. Through a combination of rewilding and adjusting income streams and subsidies, upland farms can return to being nature friendly while remaining a precious part of our cultural heritage.

So much of being concerned about the natural world involves doom and gloom, endless bad news, arguing with strangers on social media, and fears for a bleak future. Rewilding, by contrast, is positive and uplifting. In the latter stages of writing a book filled with more than its fair share of woes, it was a joy to visit a mature rewilded landscape and see its potential. I went to Knepp very early one May morning to listen to the dawn chorus and was astonished by the volume and variety of birdsong I heard. That landscape has healed and transformed itself in just two decades. We live in a sad, silent land, but it is certainly possibly to rewild it for our children.*

By now, the sun was fierce as I followed a footpath uphill across a couple of fields. I passed through a strip of woodland and emerged at the head of the charming valley. I stopped in astonishment.

'Yes!' I declared to myself, grinning. 'This is it!'

The valley opening out before me ran uninterrupted for about two miles, an enormous expanse by local standards. It was a steep vista filled with high hedgerows, trees, meadows, and a fantastic large pond. I

**Here are a few inspiring book suggestions about rewilding.* Wilding: The Return of Nature to a British Farm, Feral, Land Healer, English Pastoral, Wild Fell, *and* The Book of Wilding. *They are optimistic, encouraging reads that show it is entirely possible to feed the country, while at the same time enhancing nature rather than wrecking it.*

could see only one house, nestled in the trees near the foot of the valley.

'This is the most beautiful spot on my map,' I told myself. 'I have found the treasure.'

I stared at the view in delight, absorbing the wild landscape and feeling moved by it in a way that I had missed this year. Wildness, silence, water, contour lines and trees: this valley had everything I love, and I felt an immediate connection.

I was still taking it all in when two dog walkers appeared behind me. Aware that I was concentrating on the view, one of them stopped and spoke.

'Such a shame, isn't it?'

'What do you mean?' I asked in surprise.

'The golf course,' he explained. 'All this was a golf course until it closed seven years ago. Look, you can still see a water hazard down there.'

I expressed amazement that this wild landscape had been a manicured golf course so recently.

'Yes, it's such a shame that it's gone. Look at it now, there's nothing here. Just nature.'

OCTOBER

CONKERS

'There is just as much beauty visible to us in the landscape as we are prepared to appreciate, not a grain more. The actual objects which one man will see from a particular hilltop are just as different from those which another will see as the beholders are different.'

Henry David Thoreau

As autumn approached, I was particularly looking forward to finding conkers. Horse chestnut trees and their appealing, polished seeds are a surefire declaration of the season. The trees were introduced into Britain from the Balkans in the 16th century. They're not common in wild woodland, but are staples of towns, parks and villages. Insects gorge on their flamboyant candelabra flowers, and caterpillars feast on the leaves. Blue tits enjoy the caterpillars, and deer eat the conkers. And me? I hoard them.

As I edit this book today, months away from conker season, I have handfuls of them dotted over my desk. It might be their smooth, shiny tactility, or it may just be decades of ingrained habit, but I cannot walk past a horse chestnut tree in autumn without stooping to pop a couple more conkers into my pocket. When my pockets are full, I fling them into hedgerows to try to help them spread. I keep one in my jacket throughout the winter, rolling it round and round in my hand as I sit on crowded Underground trains or wait for meetings.

As a boy, I used to enjoy playing the game of conkers, threading a shoelace through a skewered hole then trying to smash a rival with precise but powerful swings. But I never aspired to the heady heights of the World Conker Championships. The competitive game (sport?) of conkers began on the Isle of Wight in 1848, but the inaugural World Championships didn't take place until 1965, on Ashton village green in Northamptonshire, a peaceful spot surrounded by horse chestnut trees.

The tournament began, like many excellent ideas, in a pub, in response to a cancelled fishing expedition. As with most sports invented in Britain, the rest of the world soon caught up and overtook us. Mexican Jorge Ramirez became champion in 1976, and the first overseas ladies champion was Selma Becker from Austria in 2000.*

The horse chestnut tree I stood beneath today had prematurely brown and crispy leaves. It was a victim of the horse chestnut leaf miner moth. The tiny moths, brown with white sergeant chevrons on their wings, each lay up to 180 eggs on the leaves. The larvae then bore through the leaves, reducing the health of the tree and making it more susceptible to diseases such as bleeding canker.

The temperature had fallen since last week. My fingers felt cold and it wouldn't be long until I was back wearing a hat and gloves again. The blackberries in the hedgerows were now withered and past their best.

Pretty flashes of colour had caught my eye three times this morning. The first helium balloon was snagged in a hedge. The second, a party balloon with curly ribbons, had been trodden into a field. The third was tangled in telephone wires overhead. In 2013, a primary school in England released 300 helium balloons for a project. A few made it to

**Veteran players will know there are many ways of hardening your conker before battle, though these are all banned by the official championships. Over the years, people have soaked or boiled their conkers in vinegar, salt water or paraffin. You can bake them, coat the conker with nail varnish, or fill it with glue. Twice World Conker Champion Charlie Bray says, 'There are many underhanded ways of making your conker harder. The best is to pass it through a pig. The conker will harden by soaking in its stomach juices. Then you search through the pig's waste to find the conker.' My less-committed preference as a boy was threading dozens of conkers on long rugby bootlaces, then hanging them for months in my mum's airing cupboard.*

such places as Denmark and the Netherlands, while one flew 10,000 miles to Australia. When the balloons land, they can endanger birds, fish or turtles that eat the debris or get tangled in the ribbons.

On top of the littering, a Cambridge University chemist has called for a ban on helium balloons on the basis of them being a ridiculous waste of a precious element. David Ward says, 'I will not be happy if I cannot have a medical scan in my seventies because we wasted helium on party balloons while I was in my thirties.'

I crossed a busy road and cycled down a bumpy farm track with grass growing up the middle. Looking over the square's semi-rural landscape, I was struck by the number of electricity pylons and telegraph poles. We don't really notice them, but a visitor from the past would be taken aback by their prevalence. According to the Telegraph Pole Society (life membership, £9.99), 'there is no reason why a properly treated pole shouldn't last 100 years. We know of one that was "planted" in 1908 and is still not even classed as decayed.'

A caravan in a field was being engulfed by brambles and elder bushes. A footpath led round the back of a row of sheds into a development of quiet bungalows backed up against more fields. A farmer in his tractor raised a hand in greeting as I continued around a cluster of Scots pine trees that looked incongruous in this environment, and then on past the clanging assembly of a wedding marquee.

The path ran alongside a trickle of a stream, its gentle tinkling a rarity on this map's porous, chalky landscape. I was heading towards a square church tower and the steep, red-tiled roofs and chimney stacks of an old village. A rusty sign by the church declared, 'Prohibited, all vehicles except handcarts, perambulators, invalid carriages and pedal cycles pushed by hand.'

A quiet prayer service was underway in a side chapel. It was rare for me to encounter anyone in the churches I'd visited this year. I looked at the stained-glass windows, donated 'in thankfulness for the beauty of the ever-changing seasons', and thought about the cycle of seasons I'd seen on my map. I had grown fonder of where I live with every grid square I got to know, as well as becoming better attuned to living according to the moods and pace of each month.

More than anything, my map had offered me fresh perspectives. If the constraints of real life prevent us hitting the road, we can still be explorers if we choose to be. Though I don't live in the mountains I miss, I could always deepen my connection to the land I do live on, which may also soothe my urge to be on the move. What had I discovered this year, wandering and wondering around my map for hours on end?

Some things are obvious: it rains a lot in England. The winter is very long and gloomy, and my general levels of default contentment are higher between April and October. Spending regular, scheduled time outdoors is good for the soul, wherever I go and whatever the weather. Indeed, venturing out in terrible weather increases the joy you feel afterwards. Anyone can enjoy a bike ride on a sunny day, but only the crazy folk who head out in lashing rain earn the full premium experience as we thaw out afterwards over tea and toast.

In exploring this unremarkable corner of the world, I had found that there was in fact much to remark upon. I realised that the limitations of my map were more a reflection of my own perceptions, prejudices or lack of curiosity. Every map is worth exploring, and every map is worth saving. In doing so, you might just save yourself too.

I stroked a dignified cat in the churchyard and then continued into the village. An obsolete red phone box stood on the village green, a relic of past technology. After the first standard British Post Office public phone box, Kiosk No. 1, appeared in 1921, they spread to virtually every community, just as telegraph poles and mobile phone masts have done and electric vehicle charging points need to do.

Although Giles Gilbert Scott's bright-red K2 telephone box, introduced in 1924, became a British icon, it was not universally welcomed at first, prompting a muted grey edition for beauty spots. But in the way that today's outrage is tomorrow's accepted norm, some of those places that retained their old phone boxes have now painted them red – 'Currant Red' to be precise, as defined by British Standard BS381C-Red539.

Spiky sweet chestnut husks carpeted the ground outside the phone box. A massive Alsatian dog ran at me, barking fiercely and making me jump. Its owner chuckled by way of token apology. I rode on past the burnt remains of a joyrider's motorbike, up a slight rise that offered views across farmland strung with pylons, towards the marshes and the wide river, busy with barges.

A man in the distance held the flag on a golf green while his opponent putted. Crows landed on a field of harvested stubble. The golfer missed his shot and raised his eyes to the heavens in frustration.

It took a few moments to get my bearings as I looked over this landscape, matching what I was seeing with various grid squares I'd visited over the year. Things always looked different from each fresh perspective, and places changed so much from season to season.

I rolled the smooth new conkers around my pocket, then climbed back onto my bike to cycle home and line them up along the desk in my shed. They would calm me down when I was doing things like trying to reboot the Wi-Fi router, and would keep me company through the winter months to come, with memories of good times and anticipation for the next rolling around of the seasons.

MUSHROOMS

'Stopping often, watching closely, listening carefully.'

Robert Lloyd Praeger

I began today's grid square outside the Duke of Wellington pub, which dated from 1516, two and a half centuries before Old Nosey was born. I thought about all the brawls and laughs it had seen, and the tall tales told by 500 years of drinkers. I pondered also when they'd installed a *petanque* court in the garden, a game surely more suited to Napoleon than Wellington.

The Duke of Wellington was one of Britain's greatest military heroes, as well as a Prime Minister. Although he was born way back in 1769, he lived long enough to have his photograph taken, which is impressive considering he was involved in sixty battles. And he is also a legend in the very diverse worlds of rubber boots and beef cooked in pastry.

Before the modern age of drab but pragmatic camouflage, regimental uniforms were designed to dazzle, impress, and tempt bold young men to sign up. Arthur Wellesley, as he was known before receiving his Dukedom, commissioned his shoemaker to design a boot that would be both comfortable for riding and stylish when worn with tight trousers.

As Wellington's fame on the battlefield grew, so did his status as a fashion icon, and the popularity of wellingtons has never looked back. The rubber version of the boots proliferated when millions were made for soldiers in the First World War to prevent trench foot, and demobbed men then continued wearing them for digging their gardens.

Meanwhile, celebrity chef Gordon Ramsay declared that 'Beef Wellington has to be the ultimate indulgence, it's one of my all-time favourite main courses and it would definitely be on my last supper menu.' It was also President Nixon's favourite meal (if he was telling the truth). The dish was created to celebrate Wellington's victory at the Battle of Waterloo. What better way to immortalise a national hero than by wrapping some meat in pastry? Cornish miners had already been eating pasties since the 14th century, and some sniff that beef Wellington is little more than a stolen, rebranded French *filet de boeuf en croûte*.

A quick internet search couldn't tell me what the pub had originally been called, or when it was renamed, so I climbed onto my bike and set off to discover what I'd find on the other side of the next hill or highway.

Although it was mid-October now, the sun was warm again and I was back in short sleeves. The leaves on the trees were still much greener than I mentally associate with October. Ecologist Tim Sparks noticed how green the trees were on Remembrance Sunday, so he dived back through a century of newspaper coverage to gather photos of previous Cenotaph events. At the early services in the 1920s, the tree branches were skeletal and bare, and poor Queen Mary was hunkered deep into her furs. But from the 1980s onwards, London's trees were much greener during the ceremony. It was a clever visual demonstration of our disrupted climate.

I pedalled first of all around a new housing development with smart homes and shiny, powerful cars. I passed a Range Rover, Land Rover, Audi, GTR, BMW and a Mercedes before I reached a house with a more modest Toyota. A Virginia creeper blazed blood red up a fence, some houses already had Halloween wreaths on their doors, and I was amused to see gnomes featuring heavily in one garden.*

The developers had set aside an open space behind the houses as a recreation area, with gravel footpaths and young saplings in plastic

tubes. I liberated an apple from a waist-high tree and crunched it as I rode. A small pond had been fenced off and surrounded with 'Danger' signs, as is the custom in these parts. I understand, of course, the need to keep children safe, but perhaps a compromise could be an access stile and more nuanced signs balancing the hazards of water with the joys of wildlife and the fun of poking around in ponds.

Robins sang loudly in the sunny recreation area, but the motorway's roar was even louder. Living by main roads will be quieter and healthier once we replace our shiny, powerful cars with electric vehicles. I'm excited about the prospect of quiet cars running on clean energy, and a decrease in the 7 million people, according to WHO estimates, who die annually worldwide from exposure to air pollution, including more than 30,000 in Britain.

I followed a path to a footbridge and crossed over the motorway into farmland. The smell of ploughed earth alongside the roar and fumes was discordant. A windfall of shiny sweet chestnuts and the

[Left] The renowned gardener Sir Charles Isham imported some terracotta garden gnomes from Nuremberg in 1847 and started a surprisingly enduring gardening trend. Germany had a long history of gnomes, reputed to be jolly, mischievous souls that help in the garden late at night and keep an eye on your property.
Long before that, Emperor Hadrian had real-life hermits living in his villa. The fashion resurfaced when rich English landowners began hiring people to be 'ornamental hermits'. They had to live in basic outbuildings and couldn't speak or wash. They wore rags and let their hair, beard and nails grow long. It sounds like a humiliating career, if not particularly taxing, and an even odder fashion.

fluffy remains of a fox's pigeon dinner lay on the edge of a dappled slice of trees crimped between the road and the ploughed fields. Spiders' webs strung between trees brushed against my face and I hastily wiped them off. Halloween was approaching.

The footpath took me to a 12th-century church where I sat on a tomb to enjoy the unseasonal warmth. Hop vines grew over the church windows and tendrils wound heavenward up the copper lightning conductor. Most of the woodland this church lay in had been felled for yet another new housing estate, this one covering almost a quarter of the grid square.

Riding around its streets, I felt my enthusiasm flagging in the face of the relentless reduction of our wild places. I missed being in the middle of nowhere, far from the nearest road or town. To be fair, this community seemed pleasant, which presumably is why people chose to live here. Not everyone wants a cabin in the mountains or a shack by the sea.

I listened to the squeaks and fizzes and pops of starlings chattering in the bushes by the playground. The sounds reminded me of the romance and mystery of tuning a shortwave radio in the middle of the ocean at night, searching for a signal from thousands of miles away. Hoping for a connection when you're so far from the nearest road or town, and dreaming of being back safe in the heart of a community again.

Pushing up through every grass verge around the new streets were weird and wonderful mushrooms, mysterious denizens of a truly wild and tangled world. Hundreds of mica cap mushrooms emerged from a rotting tree stump, pale brown and clustered together. Like many fungi, these saprotrophs were hard at work decomposing the dead wood and recycling its nutrients.

I found another cluster of pleated inkcaps, or little Japanese umbrellas, a delicate fungus that appears overnight on grassland after rain. They appear, grow, release their tiny spores and then decay, all within twenty-four hours, leaving no trace of ever being there.

Also pushing through the grass were shaggy inkcaps, or lawyer's wig mushrooms. After depositing their spores, or after being picked, they turn black and autodigest (dissolve themselves) in just a few hours. They are tasty, but you have to cook them soon after harvesting before

they disappear altogether, leaving only a black stain. It is small wonder that, in olden times, people were filled with a sense of awe and the supernatural. Indeed, the more we learn about fungi, the more mysterious and remarkable they become. They break down pollutants, create soil, and are vital to life on earth.

I got up off my hands and knees, waved goodbye to the mushrooms and pedalled homewards. I was going to miss these weekly excursions. The Duke of Wellington once said the 'art of war consists of guessing at what is on the other side of the hill', and that continual mystery had been one of the pleasures of exploring my map this year.

I have shifted from pursuing adventures of a lifetime towards a lifetime of just trying to live a little adventurously every day. Moving from intercontinental journeys to one map and a microscopic inspection of my world had required me to develop new habits and to try to leave behind much of the noise that crowds my mind all the time. Thoreau called this approach his 'village mind', recognising that it required discipline to go to the woods in order to do less, but to do that well.

LEGACY

'Most of us are still related to our native fields as the navigator to undiscovered islands in the sea. We can any afternoon discover a new fruit there which will surprise us by its beauty or sweetness. So long as I saw in my walks one or two kinds of berries whose names I did not know, the proportion of the unknown seemed indefinitely, if not infinitely, great.'

Henry David Thoreau

It was a morning of fresh sunshine and a chilly breeze, that day defined in *The Meaning of Liff* as 'Brithdir – The first day of the winter on which your breath condenses in the air.' There had been the first faint frost as I pedalled out this morning, pulling on thick gloves and feeling the pinch of cold on my nose. The year was drawing down. The season's early fieldfares flew over the fields, a flight pattern of several wing beats, then a quick glide, eager to forage on the abundant hawthorn berries. Fieldfares look like thrushes but stand taller, move in big hops, and spend the winter in flocks of hundreds.

Reading about fieldfares led me down a Twitter rabbit hole via the #vismig hashtag, of which I'd never heard. Visible migration (which I'd never heard of either) is the 'visible' migration of birds and butterflies during daylight. Many other species migrate at night (#nocmig), which is harder to monitor unless flocks reach the coast at dawn, an event known as 'falls'. We learnt a lot about nocturnal migration when radar was invented in the First World War. All those birds could now be observed for the first time, showing up on radar screens as 'phantoms' or 'angels' flying through the dark skies in silent flocks.

I climbed a steep hillside to enjoy a misty, pale view westwards over miles of woodland and villages. I rested on a bench, poured a cup of coffee from my flask, and gazed out over a landscape that felt far more like home than it had done twelve months ago.

A plaque on the bench commemorated an old man 'who lived in this village and found peace in these hills'. A photograph of a young man who had died recently, aged just twenty-one, was also pinned to the seat. I searched online for his story, but, unusually, I couldn't find any information about him. The very searchability of *everything* had added so much to my project. I had learnt a lot by stravaiging around the map getting my boots muddy, but I discovered even more from all the googling prompted by each outing, following my curiosity around the internet.

The hillside to my right was covered in trees, except for a single open field that had been cleared from the wood, like a bald patch of buzzcut on someone's scalp. The woodland was a mosaic of green leaves, autumn colours and dark evergreens. When I entered the wood, the ground was strewn with colourful leaves and countless beechnuts. A huge tree had thick straps bound between its primary branches to prevent the top-heavy tree splitting in half in heavy winds.

I crunched through the wood to another open hilltop, which offered more wonderful views across the still-misty lowlands, though up here the sun was now bright and warm. I passed a large cross mounted on the hilltop and two horses, one grumpy, one friendly. The friendly horse was adamant I should give him a stroke before continuing.

A bench on the hill had a solar-powered speaker that gave information about the local area when you pressed a button. A bunch of lost keys had been hung on a fence, a beam of sunshine shone right through the hollow trunk of a gnarled ash tree, and a passenger jet flew high overhead, chalking the flawless blue sky with its contrails. I used to fly often, and far, and I enjoyed it very much. But I hadn't travelled anywhere by plane for a few years now, feeling unable to justify it anymore. I missed the excitement and the variety of fast foreign travel. But I missed it less than I thought I would. I miss it less the more time goes by. And I certainly missed it less once I poured myself into the surprising richness of finding a world on one map.

I have always revelled in journeys through wild places, treasuring the awe of mighty landscapes and the pleasant pleasures of clear rivers and forests filled with birdsong. Nature has been a beautiful backdrop to the challenges of all my expeditions. But, to be honest, I never paid close attention to it. I was a fan of nature, of course, but it was an abstract relationship. I liked sitting outside my tent watching the sun go down, but I didn't ponder what used to be here a hundred years ago, consider whether the wildlife was less or more abundant than a generation ago, or learn the names of local flowers.

This year, on the other hand, I had discovered more about nature than ever before. I had become engrossed by marginal landscapes such as the crumbling factories with swallows nesting in the eaves. I appreciated the silted-up canal, lively with dragonflies, and the spring leaves on the sycamore tree by the kebab house. I had found solitude in abandoned buildings, and beauty on neglected marshes. It had been a fantastic project.

Sharing my magnificent hilltop view was a grand old manor house that is now a venue for Christian conferences and retreats. A noticeboard on the outer wall offered 'Evidence for the Resurrection', although evidence that depends on faith pushes my credulity these days. Behind the manor, the land flattened out into grassy fields dotted

with ancient oaks. I walked across the fields towards a wood, on paths that were becoming slippery and muddy again as winter approached.

A green woodpecker flew between two trees, and a flock of wood pigeons flapped noisily up from the ground. I smiled to myself. I had seen loads of pigeons all year, on literally every grid square, and yet had never bothered to write about them. I had grown very fond of birds this year, finding it satisfying to watch them, listen to them, and allow them to help me become more observant, but pigeons didn't interest me at all. I think this was because of their lack of mystery as you see them everywhere, their dumb cooing of 'my toe hurts, Betty, my toe hurts, Betty', their clattering take-offs, their scruffy nests, and by association with their tough urban cousin, the feral pigeons, who flap and strut and shit in such numbers across our cities.

Yet this dismissive discrimination was not in keeping with a project that urged me to look for beauty and interest in every corner. I had been addressing my apathy by reading an interesting book by Jon Day called *Homing: On Pigeons, Dwellings and Why We Return*. We have domesticated pigeons for thousands of years, even holding the bird as a symbol of peace. But these days, many people consider them vermin because of their dominance of cities and the mess they make. Day's book was a reminder for me of their beauty and incredible homing abilities.

Our wood pigeon population has exploded by 170 percent in the past fifty years, at the same time as another cousin, the turtle dove, has become the fastest-declining bird in the UK and our bird numbers overall have dropped by almost a third. Pigeons now have nearly the same total biomass as all of our other songbirds put together. Their ability to live almost anywhere, including among humans, means they thrive in barren, denatured, simplified landscapes. I don't blame the pigeon for its success, but its dominance highlighted how relatively little of everything else there was on my map.

I found a mighty beech tree in the woods, hundreds of years old and looking awesome. The circle of bare earth at its base mirrored the enormous span of its canopy, showing how effectively it filtered out the sunlight and eliminated the competition of any young upstarts below. Nailed to the bole of the tree was a small plaque saying, simply, 'Peter 1925-2009'. The lack of biography left me contemplating how little we will know of Peter's eighty-four years (or mine, or yours) once the memories of whoever knew him and fixed that plaque have faded.

An old tree on a sunny morning is a fine reminder of one's place in the order of things, a marker of how wise it is to savour today rather than worrying about intangibles such as our legacy, or our desire to hustle, to chase, to 'succeed'. Perhaps the best impact we can leave behind is, as the old adage has it, to plant trees (and metaphorical trees) whose shade we shall never sit beneath.

Or, perhaps, I might settle for the legacy of 'our Fred', whose sixty years of life were commemorated on a charming carved bench in those same woods, along with the inscription that 'you loved to walk through these woods on your way to the pub for a glass of lunch'.

A glass of lunch with friends, plus some planted trees, would do me very well as an epitaph to work towards.

PARAKEETS

'I think that by retaining one's childhood love of such things as trees, fishes, butterflies and toads, one makes a peaceful and decent future a little more probable.'

George Orwell

I cycled to a small town that I knew as a motorway junction and a monstrous snarl of a roundabout. And yet I was riding towards it down pretty lanes fringed with red and yellow leaves that swirled and spun in the wind. It was disorientating not to have thought of this place in this way before. What would I discover on the last of my fifty-two grid squares?

I had spent an entire year on a small map that I'd feared would be boring and meagre. But I saw now that I was nowhere near to knowing it fully. I would need to continue at the same pace for another seven years before I even visited every square, let alone travelled around each one in each season, during rush hour or at dawn, by bike or on foot, alone or with a companion. You never pass through the same grid square twice. I can never know even one map, not in all its seasons and weather, nor all its harvests and wildlife. And I had barely begun on the countless human stories and history intertwined in my nondescript neighbourhood.

I felt sad that my year of exploring this area was ending, though of course there was no need for it to stop. I hope that my new habits of curiosity will continue for ever. You can see the particular or the universal everywhere, depending on how you look. There is a world in a wild flower, an eternity in an hour.

It was serendipitous that I passed a house called Galleons Lap with two gigantic gnomes (surely an oxymoron) and a Lego dinosaur in the front window. Looking up the unusual name, I discovered it came from *The House at Pooh Corner*, where Pooh and Christopher Robin 'walked on, thinking of This and That, and by-and-by they came to an enchanted place on the very top of the forest called Galleons Lap… Sitting there they could see the whole world spread out until it reached the sky, and whatever there was all the world over was with them in Galleons Lap.'

This book is distinct from many I've written as it's not about voyaging from here to there. Indeed, it's a book about going nowhere much and doing not a lot.* I'd had serious reservations at the start about dedicating a whole year to just twenty kilometres. To my relief, it had turned into one of the most stimulating journeys of my life. The most important step in the whole experience was merely deciding that my local map was worthy of proper investigation. The expectations we look with are the greatest restriction to seeing anything interesting.

When I travel in foreign countries, I'm desperate for eventful things

**Well done for making it this far.*

to happen. I peer from bus windows, hungry to absorb every detail. I honour the taxi driver as an oracle of wisdom. I delight in exchanging a few words with a shopkeeper. I am determined that this day will be memorable, photogenic and informative. And, thanks to a combination of both expectation and observation, I am usually right. Yes, those places are fresh to me, but a lot of their magic also stems from simply bothering to look. Back home, by contrast, everything is familiar, so I have low expectations about what I'll see, and the results are correspondingly humdrum. It's like looking out with disinterest through grimy spectacles.

The challenge I'd set myself this year was to be as excited and intrigued about home as I was each time my passport was stamped in a new country. I had worried that after years of global adventures this small map would make me feel claustrophobic. But I was wrong. It was a year of constraints, yes, but adventure with constraints is not only more responsible to the planet, it also forces you to be more imaginative. If you find somewhere new a few miles from your home, then you're exploring the world just as much as someone crossing the Empty Quarter desert in Arabia.

I turned down a byway that tunnelled through the undergrowth, with paths forking off like streams as I made my way to the town church. Parakeet screeches rang out from the church's yew trees. They are an increasingly common bird around here, but not one that I like. Neon-green parakeets rocket over my shed every afternoon, flying fast with a tropical squawk that sounds jarring in England. How they came to be thriving here is the source of excellent urban legends: the Great Storm of 1987 damaged aviaries; a plane crashed through an aviary roof in Syon Park in the 1970s; Jimi Hendrix released a pair in the 1960s; they escaped from a Humphrey Bogart and Katharine Hepburn film set in the 1950s…

But boring old science and forensic analysis suggests that the population has just grown from the odd pet owner releasing a bird here and there, either accidentally or perhaps in panicked response to occasional newspaper frenzies about the risks of 'parrot fever'.

'Squawk!' screeched the colourful parakeets. 'Squawk! We are here because you were there!'

At some point in the past, we went to Africa and India, messed about, and brought some parakeets home because it suited us. They settled, thrived, and now they live here. What right does that then give me to complain, to chunter my lopsided perceptions of what ought to belong in 'my' country? Very little. Yes, they are colourful and noisy and not typically 'British', but perhaps we could do with a bit more colour and volume in our lives.

I meandered around the churchyard for a while, before being drawn towards a grave covered in plastic flowers and heart-shaped helium balloons. A large photograph showed a lad posing in boxing gear with his fists up in a guard position. Traditionally, people from the Romany and Irish travelling communities buried photographs of the deceased, but that has evolved to placing photos on the graves or incorporating them into the gravestones.

Taking out my phone, I learned the sad story of this keen young boxer from the Traveller community, whose last fight was with bone cancer. Click led on to click and I found myself, many minutes later, still scrolling sadly through an Instagram tribute page to the boy, filled with cheerful photos from his short life and then brave thumbs-up images from hospital beds. I sighed, turned off my phone, and cycled away.

I rode out towards open fields of cropped grass, separated by fences and stocked with horses. There were long views under grey skies over undulating land sprinkled with modern developments and pylons. Dogs barked in the distance and the sound of kids playing carried through the air. The motorway was quiet today, though continuously audible. A school bell rang, summoning the children back to work. It felt satisfying to look around for the last time and register how the places I'd been to this year fitted together.

'OK, so there's that big bridge… Oh yeah, there's the huge pylon. I see where I am now.'

The hedgerow brimmed with the reddened leaves of spindle trees with their orange berries in pink capsules. The presence of spindle often shows you're in ancient woodland, or where it once stood. After being baked and powdered, the spindle's inedible, laxative fruits were traditionally used to treat head lice in people and mange in cows. The pale wood is hard and was used to make spindles for spinning wool, hence its name. Today it is mostly used for artists' charcoal.

Pedalling on down a lane, I passed some tiny bungalows, then an enormous gabled house with ornate fences, a gravel turning circle and a swimming pool. We all live such varied lives, even along the same road, on the same square, on the same little map.

A wren let loose a burst of its distinctive staccato song, so familiar to me by now, as I paused on a narrow bridge to watch a train chug beneath me. I peered to look at a snail creeping along a wall, one of around 120 species in Britain that range in size from a single millimetre up to the Roman snail, which is the size of a golf ball and was brought to Britain by the Romans.

I turned onto a footpath through a brown, empty field dotted with fly-tipping. Someone, at some point, had hurled a vacuum cleaner high into an oak tree, and there it dangled still, tantalising me with its mystery. Who? Why? How? When? There are a million stories everywhere.

I was drawn, one last time, to the spaces that are hemmed in by infrastructure such as motorways, railways and factories. Perhaps they feel exotic to me because they are so at odds with my upbringing of hills, moors, and miles and miles of space to roam free. But these noisy, semi-feral spaces have fascinated me all year.

For the final time then, I enjoyed exploring a wood that was pressed up against the power and speed of a motorway. I peered through the branches at lorries hammering past just metres below me. A jay flew by with an acorn in its beak, its electric-blue feathers gleaming even in this dark wood.

Years of old rope swings, snapped or frayed and out of reach, dangled from the limbs of a tall tree. The latest swing was a dark blue rope with a thick stick tied to it. I had a go, which turned into about ten, launching out over a steep and muddy slope. It was a perfect spot for kids to slither up and run down, arms flailing, eyes wide, shrieking with laughter. I spent a lot of my childhood messing about, unsupervised, in woods like these, and I value those old memories. I was glad that some children today were doing the same. I was grateful, too, for all the new memories I'd accumulated exploring this map.

I couldn't think of a more apt way to finish off the grid square and the year than by barrelling down a muddy single-track path I never knew existed, on a folding bicycle designed for city streets, through a little wood carpeted with golden leaves and discarded drinks cans, alongside a noisy motorway.

I freewheeled down out of the wood and back towards the road. A workman stopped his digger as I drew near. He climbed down, eager for a chat and to tell me all about the driveway he was constructing.

Then he asked me, 'Where does that path go up there? I've never been.'

'Why don't you go and explore?' I suggested. 'Walk up there on your lunch break.'

'You know what, I might just do that,' he replied.

SIX MONTHS LATER

As I made the final edits to this book, I realised there was one more ride I wanted to do before letting go of it. I wanted to cycle through every square of my map in a single, unbroken journey. Starting in the middle and circling outwards through all four hundred grid squares appealed to my old instinct to spiral away, faster and faster, until sling-shotting beyond the gravitational pull of home and blasting out into the world. But I decided instead to begin on the map's outer fringes and to circle inwards until I arrived home, or, more precisely, reached the pub. It fitted better with my search for somewhere to belong, and was a dive towards the heart of things.

The route I plotted through every grid square was 300 miles long, climbed 5,800 metres and looked utterly ridiculous, like a plate of wiggly spaghetti. I transcribed it from my paper map into a digital navigation app to save me having to check the map at every junction.* Two-thirds of it was on paved roads, the rest was byways, footpaths and the odd spot of gentle trespassing. I didn't know where I would eat or sleep, but I was confident those details would work themselves out over the next four days. It was time for 300 miles of adventure on

*RideWithGPS *worked brilliantly.*

a twenty-kilometre map (I think in miles when I'm cycling, even on a metric map).

I felt the old combination of joy and nerves as I ate a large breakfast, squeezed a last-minute extra warm top into my pack, and hit the road. Spring was on its way as I pedalled through familiar grid squares towards the edge of my map, the roads lined with blackthorn blossom and the first blush of hawthorn leaves. This was my world: my roads, my sunshine, my plastic junk floating in the river, my broken bottles. It was my home, and I was happy to be here.

I crossed paths with a cyclist who was out from the city for a day ride. He drove buses for a living, wore designer cycling clothes and was jealous that I was going to camp out tonight.

'I'd love to do that, but I'm too scared,' he said, pointing at the bags on my bike. 'I've camped with friends, but never on my own. I need to lose my camping cherry.'

I asked why he'd ridden out this way.

'I like scuzzy stuff. Old, broken, weird, forgotten. That sort of stuff, you know? Scuzzy. There's loads of it round here.'

He looked embarrassed at this odd admission, but I knew exactly how he felt.

'Me too!' I said. 'Edgelands, bus depots, graffiti. Scuzzy! I love all that too.'

I certainly got my fill of scuzzy on this ride. But I also heard the first lapwings and skylarks of the season. And while riding the hard shoulder on an unavoidable but stressful stretch of dual carriageway, I spotted my first brimstone butterfly of the spring, a flash of yellow darting among the roadside litter.

Around mid-afternoon on the first day, I paused for coffee and rocky road cake in a posh village of black-and-white Tudor buildings and Teslas. I flicked idly through the *Financial Times Weekend* supplement that a previous customer had left behind. I was embarrassed by how knackered I already felt. All those weekly bimblings clearly hadn't helped my endurance. Earlier, I'd rested in a rundown town that smelt of weed. A Home Office Immigration Enforcement van pulled up, sirens wailing, and the officers piled swiftly into a block of flats. I watched a street cleaner sweep a cigarette butt, miss, sweep again, miss, then shrug and walk on.

My map was actually far more rural than I had realised, more wooded and hilly, and less built-up. That was a nice realisation, but it also meant I failed to find a village pub to eat dinner in, so had to backtrack a few miles to a McDonald's in a retail park. I whiled away a couple of hours there, merrily stuffing my face with delicious Diacetyl Tartaric Acid Ester of Mono- and Diglycerides while I charged my camera batteries and stayed warm until it was time to sleep.

I slung my hammock beneath a three-quarter moon in a wood I'd visited earlier in the year. The night was bitterly cold, and by dawn my water bottles had frozen solid. I cursed at the stupidity of shivering through a long night just a few miles from my cosy bed. At first light I pedalled quickly to the nearest petrol station to recover with hot bad

coffee. I resolved to detour home later for a warmer sleeping bag and a tent.

I was weary on day two, slogging up muddy trails through woods and down slippery footpaths across fields. Heading through every grid square like this was making my map feel huge. Although I'd often felt concerned this year about the scale of new housing developments, this microadventure showed me there was enough space for everyone here, as well as for wildlife and farming.

We can certainly rewild some land for nature once people are enthused to think it's a valuable idea. Soil can heal and hedgerows and ponds can recover once there is a demand to support sustainable, regenerative farming. The river filled with slimy lager cans is a clear chalk stream at heart, home to water crowfoot flowers and otters, and just waiting to revitalise as soon as the political will is there to get it cleaned up. This is all so achievable, making it an exciting time to be an engaged adult concerned with leaving the world in a better state for our children.

Each day of my ride, the trees seemed to turn a little greener. The dawn chorus woke me earlier each morning. Winter was on its way out. Brighter days lay ahead.

'Here we go again,' my map said to me. 'Another season, another lap of the sun, another lap of the map. Another chance to make the best of things and choose to bloom as brightly as you dare.'

My sinuous route wound round and round, nonsensically but pleasingly. It was astonishing how many new back roads and bridleways I found, even from just whizzing straight through grid squares. Before this year, I had focused too much on the oceans and mountains I did *not* have where I lived. But that had been replaced by a strong sense of how much this map *did* have to offer. I was proud of how much I now knew about this area. If you joined me for a ride, I could show you around well.

I passed rambling old houses with tennis courts and trampolines,* decaying tower blocks with washing lines and barbed wire on the roofs, and repeating suburban streets with everything in between. As I rode

**The sort I will buy once I start writing blockbuster bestsellers, rather than niche books about bicycling around business parks.*

randomly through so many diverse lives, I hoped that contentment was spread more evenly across my map than wealth or opportunity seemed to be.

I slept well in my warm tent that night, back into the groove of the cycle touring life I knew so well. This trip was no different from all the times I'd ridden hundreds of miles in other countries, apart from the need to accept that I wasn't moving very far, and to be OK with that.

There was a lot to love about riding a long way yet never being more than ten miles from my fridge. I loved the detours down hidden-away footpaths hemmed in with chain-link fencing around railways shunting freight goods, faceless facilities management warehouses, sewage works, and hedgerows scattered with nitrous oxide canisters.

I loved the estuary with its derelict broken windows and its birdsong and dramatic skies scudding with storm clouds, bursts of rain, and rainbows. I loved the marshes, pylons and dilapidated wrecks of boats hauled up on muddy creeks, paint fading, wood peeling, full of intrigue and character.

I loved the peaceful hideaways that few people knew about, the silent valleys and my first bluebells of the year. I loved the wild predators who made this tame landscape their home: the foxes, the marsh harriers, the dragonflies and brown trout.

I loved the ancient churches and gigantic trees standing proud from long before the industrial and agricultural revolutions, and I loved knowing that they will still be here once we get back to living in harmony with nature and our communities again. And I loved that I was free to enjoy all these places on miles of public footpaths.

I didn't much like all the 'Keep Out' signs, the locked-away lakes, the litter, the dog poo bags, and the miles of new streets named after the meadows, birds and trees they had replaced. But even so, there was still plenty of space for me to watch the sun set behind a titanic oak, the golden sky splintered like a shattered windscreen by the tree's thousands of crooked twigs and branches. As the first stars appeared, I pitched my tent in an empty wood that I shared only with a duetting pair of tawny owls.

My frustration at the start of this book about living in a built-up area felt mistargeted on this ride. It wasn't that there were no woods to explore or trails to ride here. My real problem was having nobody to do those things with, and I hadn't solved that problem this year.

I frequented plenty of cafés and pub gardens, delighted to squander more money on cold lagers and steaming Americanos in four days than I had during four years of riding round the world on a young man's frugal budget. I savoured the lone traveller's prerogative of listening in anonymously from the edges, overhearing café conversations about a long-lost brother from Australia, one half of a phone argument with a boyfriend, and a place to buy excellent belly pork. There are so many lives playing out on this small map, so many demands on the land's resources, so many differing priorities.

My meditative looping round and round the map towards its centre gave me time to think about the ways our priorities differ. Not everyone will agree with the things I care about. But if any topics in this book do resonate with you, I urge you to tell your friends about them and to write to your MP to engage them in the conversation (it is quick and easy to do via www.theyworkforyou.com). Nothing major changes without sufficient people speaking up, and without voters holding their officials to account.

Ultimately it is the government who will decide whether or not we meet our climate targets, whether we shift swiftly to electric vehicles and effective public transport, whether sustainable farming is adopted seriously, whether paying public money for public goods will result in cleaner rivers and an increase in rewilding and conservation, and how much importance is placed on helping people to spend more time exercising in the outdoors, with improved access and education on how to care for nature.

Dodging broken glass down a back alley on my final morning, I came across an abandoned building with only the walls still standing and old vehicles rusting among self-seeded sallow saplings. An elderly gentleman was walking his dog and I asked if he knew what the building had been.

'A builder's yard, I believe,' he said. 'It's been abandoned as long as I've known it though. Forty-odd years since I moved into the area.'

'Where did you come here from?' I asked.

'Only from town, just a couple of miles, you know. I don't know about you, but I'm not exactly a world traveller.'

'You must like it here to have stayed so long?'

'It's all right, mate. But they've built all those new homes on the old community field. Barry used to keep those pitches immaculate, you know. I can't believe they squashed so many houses on there. Sometimes I think it's time to move on. But there's all the upheaval. And it's what you know, right? People get used to a place. It's home, isn't it?'

I had pedalled, pushed and carried my bike for four days, going round and round my map. I fixed punctures and lifted the bike over gates. I took selfies and wrong turns. I fell off my bike, hard, and stuck my middle finger up at the passing car taking a photo of me as I lay in pain on the pavement. I ate cakes and crisps and croissants and curry. I rode through hundreds of grid squares. But in the end, I didn't manage

to ride through every square.

I had overestimated how far my unfit legs could carry me each day, and underestimated how much I like cafes these days. So I ran out of time before my family returned from a trip away. I turned for home and pedalled back to them (via the merest of stop-offs in a pub garden to toast the end of this long project).

Even after a year plus four days, I still hadn't covered many of my map's roads and footpaths. There were still grid squares I had to label as *terra incognita*. But I feel content knowing they are out there waiting for me to explore in the years ahead.

My small weekly forays had been a contrast to the bigger adventures from my past. Long, slow journeys through unfamiliar landscapes and cultures are an astonishing privilege, stuffed with lifelong lessons and perspectives. Let nothing discourage you from heaving on a rucksack or strapping a tent to a bike and leaving town. Go! You can't beat that feeling of being untethered and free. Do it if you possibly can. Those star-filled nights, huge horizons, and strangers who will become friends: they are out there waiting for you, your life lying before you like a blank page, brilliant with possibility, daring you to begin.

Yet Thoreau, whose dogged enthusiasm for staying put and paying attention makes him the unofficial ambassador for this book, boasted that he had 'travelled a good deal in Concord', his small home town in Massachusetts. I like that idea too. For if I were to boast in my local pub, pint in hand, I could certainly bore punters with memories of cycling through Albania and Zimbabwe, or paddling to Barbados or through the Yukon. But I would now also tell them, with fair confidence that nobody else knew these places, about a delicious chalk stream under a motorway bridge near here, a hilltop of summer orchids and oregano, and a twisted oak tree, six metres round and centuries old, standing dark and strong in fields of snow.

I am proud to know these familiar little spots, for they have helped me learn to appreciate where I live and feel more attached to it, despite Thoreau's insistence that a landscape can 'never become quite familiar to you', no matter how long you live there.

But can a single map really be enough exploration for a lifetime? Pootling around one map for a year rarely felt like an adventure, I'll admit. But it did often feel like exploring. I enjoyed many tingles of surprise on my map of small wonders. I won't push your credulity in claiming it was epic, but something about the experience resonated with the sliver of my soul that wants always to look beyond the horizon. My weekly meanderings did a decent job of keeping a lid on that restlessness. So much so, in fact, that I feel something akin to vertigo at contemplating the prospect of having the entire globe to explore.

If you pick up a map of your local area, choose a grid square at random, and begin walking around it with your eyes open, you'll soon be mesmerised by the possibilities for local exploration. After that, it is up to you. What will you look for? What will you care about and want to take a stand on?*

My map has changed my perception of home, made me less tempted to fly, and more motivated to care for the environment. There is so much potential for a future full of positive stories, if only we demand change and take action.

This local map could fuel my curiosity for ever, in a way I once thought only distant places could do. My map is a fractal of the world. Today is a fractal of my life. To know one place well and to make it better is the work of a lifetime. And so, yes, a single map can be enough.

**When you start exploring your own local map, please tag your photos on social media with #ASingleMap to help inspire others.*

Before you go: please rate this book!

Please consider leaving a quick review on Amazon for this book, or sharing a photo of it on social media with the hashtag #ASingleMap. These both make a massive difference to the success of a book. Thank you.

And don't forget to buy your own local map and get exploring…

RESOURCES

- Buy your own Explorer map (1:25,000). www.shop.ordnancesurvey.co.uk/custom-made
- Many other countries have similar mapping agencies to the Ordnance Survey. For example, try the 1:24,000 topo '7.5 minute' maps in the USA (www.store.usgs.gov), MyTopo for the USA and Canada (findamap.mytopo.com/findamap), the Centre for Topographic Information in Canada (www.gotrekkers.com), NATMAP in Australia (www.xnatmap.org), the Norwegian Mapping Authority (www.kartverket.no), France's Institut Géographique (www.boutique.ign.fr/cac/cac-grand-public.html), Scanmap in Denmark (www.scanmaps.dk), Calazo for Finland, Norway and Sweden (www.calazo.fi), mySwissMap in Switzerland (www.shop.swisstopo.admin.ch/en), and the IGN in Spain (www.ign.es/csw-inspire/srv/eng/main.home).pn
- While I prefer to use a paper map for planning adventures, digital versions are also useful. The Ordnance Survey app is excellent for the UK. Maps.me (www.maps.me) and OpenStreetMap (www.openstreetmap.org) are free world maps to explore wherever you live.
- RideWithGPS was invaluable when I tried to cycle through all four hundred grid squares.
- Seek identifies the plants and animals all around you. www.inaturalist.org
- Merlin listens to the birds and shows real-time suggestions for who's singing. www.merlin.allaboutbirds.org
- Star Walk shows a real-time interactive sky map on your phone. www.starwalk.space
- 1000 Hours Outside helps to match nature time with screen time. www.1000hoursoutside.com
- Geocaching is a GPS treasure-hunting game. www.geocaching.com
- Hyperlocal Weather helps you to decide when to dash for a café and when to just dance in the rain. www.hyperlocalweather.app
- UK Soil Observatory has masses of information about the ground you walk on and the ways that land is used. www.ukso.org
- Slow Ways encourages walkers to rediscover unused footpaths and engage in more leisurely walks. www.beta.slowways.org
- The Ramblers are trying to save 49,000 miles of lost paths. www.ramblers.org.uk/support-us/dont-lose-your-way

TAKE ACTION

- If any of the issues raised in this book have made you think, Take the Jump helps you to try six shifts to protect the earth and live with joy. They are clear, constructive, impactful, and doable. www.takethejump.org
- The six challenges are:
  1. *End Clutter. Keep products for at least seven years*
  2. *Travel Fresh. Avoid personal vehicles where possible*
  3. *Eat Green. A plant-based diet, no waste, a healthy amount*
  4. *Dress Retro. Three new items of clothing per year*
  5. *Holiday Local. One flight every three years*
  6. *Change the System. At least one life shift to nudge the system*
- Right to Roam is a campaign for millions more people to have easy access to open space, and the physical, mental and spiritual health benefits that this brings. www.righttoroam.org.uk
- Rewilding Britain aims to tackle the climate emergency and extinction crisis, to reconnect people with the natural world, and help communities thrive. www.rewildingbritain.org.uk
- Trash Free Trails (re)connects people with nature through the simple yet meaningful act of removing single-use pollution from wild places. www.trash-freetrails.org
- Tips on rewilding your garden: www.tinyurl.com/rewildgarden
- Pay attention to environmental and climate policies before deciding who to vote for in elections. See how your MP votes on such issues as river pollution and conservation. Send them a message to let them know what matters to you. www.theyworkforyou.com
- Nature Connectedness from the University of Derby is a free, online short course that shows how we can build a new relationship with nature for the wellbeing of both people and the rest of the natural world. www.tinyurl.com/natureconnectcourse
- Discover people, ideas and news pointing the world in a positive direction. www.theprogressnetwork.org
- These recipes helped to persuade this vegan-mocking carnivore that tasty plant-based food not only existed, but that it was easy and cheap to prepare:
  - *The Green Roasting Tin: Vegan and Vegetarian One Dish Dinners: www.tinyurl.com/greenroasting*
  - *Anna Jones' Proper Chilli: www.tinyurl.com/properchilli*
  - *Very easy dhal: www.tinyurl.com/easydhal*
  - *Sweet potato dhal: www.tinyurl.com/easydhal2*

FURTHER READING

I read well over a hundred books while researching this book. Out of all those, these had the greatest impact on me:

- *Eating Animals* by Jonathan Safran Foer
- *Feral: Rewilding the Land, Sea and Human Life* by George Monbiot
- *Ravenous: How to get ourselves and our planet into shape* by Henry Dimbleby
- *Regenesis: Feeding the World without Devouring the Planet* by George Monbiot
- *The Trespasser's Companion* by Nick Hayes
- *The Uninhabitable Earth* by David Wallace-Wells
- *Wilding: The Return of Nature to a British Farm* by Isabella Tree

ACKNOWLEDGEMENTS

A very warm Thank you to everyone who helped with this book: David Charles, Jon Doolan, Mike Sowden and Zanna Davidson for being great editors. Dan Hiscocks, Simon Edge, Rose Shepherd and the team at Eye Books for calmly bringing the book together and getting it out into the world. Nell Wood for the fantastic cover design and interior layout. Martin Hartley, Emily Garthwaite and Jim Shannon for helping whittle down 8000 photographs. Ordnance Survey for the brilliant maps that open up the countryside to so many of us. The volunteers who update Wikipedia, the niche bloggers I stumbled across, and the invaluable Our World in Data website for teaching me so much. Zac Bannister for his expertise on monkey puzzle trees, which is impressive for a nine-year-old. The volunteers who cast a beady eye over parts of the text and shared their specific expertise: Abhi Arumbakkam, Adam, Aimee Alsop, Amy Chapple, Amy-Jane Beer, Andrew Brown, Andrew Chapman, Andrew Knights, Bernice Webb, Caroline Dawson, Chris Ellison, Christopher Redwood, Cory Mortensen, Craig Humphrey, Dan Freeman, Dan Manley, Daniel, Daniel Winter, Danny Butler, Danny Child, Debs Mutton, Denise McCullagh, Eliza Ecclestone, Ethan Purdy, Ethan Trencher, Faith Huggins, Felicity Hindle, Helen Marshall, Ivan Routledge, James Nicholson, Jonathan Moses, Julie Phillips, Karl Gwilliam, Katrina Roszynski, Kelli Jackson, Layla Westwell, Leon McCarron, Lewis Winks, Lord Rob Bushby, Luke Woods, Mark T, Michael Desrosiers, Monika, Neil Heseltine, Nigel Fishburn, Peter Powell, Rachael Walshe, Rachel Schneiders, Rob Lilwall, Robert Rose, Sarah Hammond Ward, Sophia Zell, Stephen Wilkinson, Stuart Antrobus, Tim, Tim Chippindall, Tom Dauben, Tracey Gait. And runner Rickey Gates for the pithy line in his film *Of Fells and Hills*, which has always stuck with me: 'In the end I think that a single mountain range is enough exploration for an entire lifetime.'

ABOUT THE AUTHOR

Alastair Humphreys has been on expeditions all around the world, travelling through more than eighty countries by bicycle, boat and on foot.

Alastair was named as a *National Geographic* Adventurer of the Year in 2012 for his concept of microadventures – 'local adventures for everyone'. The Royal Geographical Society presented him with the Ness Award in 2023 for 'his long-standing contributions to promoting a greater understanding of our world and wider public engagement with the outdoors'.

He creates videos, podcasts, newsletters, and has written sixteen books for adults and children.

www.alastairhumphreys.com
@al_humphreys